ADVANCES IN ENVIRONMENTAL MONITORING AND ASSESSMENT

Edited by **Suriyanarayanan Sarvajayakesavalu**

Advances in Environmental Monitoring and Assessment
http://dx.doi.org/10.5772/intechopen.75847
Edited by Suriyanarayanan Sarvajayakesavalu

Contributors

Carlos Alexandre Borges Garcia, Helenice Leite Garcia, Igor Santos Silva, Maria Caroline Silva Mendonça, Sunita Verma, Divya Prakash, Manish Soni, Kirpa Ram, Meng Gao, Norma Patricia Munoz Sevilla, Mariano Norzagaray Campos, Maxime Le Bail, Mohamed El-Raey, Hamdy Abo-Taleb

First published in London, United Kingdom, 2019 by IntechOpen
IntechOpen is the global imprint of INTECHOPEN LIMITED, registered in England and Wales, registration number: 11086078, The Shard, 25th floor, 32 London Bridge Street
London, SE19SG – United Kingdom
Printed in Croatia

British Library Cataloguing-in-Publication Data
A catalogue record for this book is available from the British Library

Additional hard and PDF copies can be obtained from orders@intechopen.com

Advances in Environmental Monitoring and Assessment, Edited by Suriyanarayanan Sarvajayakesavalu
p. cm.
Print ISBN 978-1-83881-009-2
Online ISBN 978-1-83881-010-8
eBook (PDF) ISBN 978-1-83881-011-5

For EU product safety concerns:
IN TECH d.o.o., Prolaz Marije Krucifikse Kozulić 3, 51000 Rijeka, Croatia,
info@intechopen.com or visit our website at intechopen.com.

Meet the editor

Dr. S. Suriyanarayanan, MSc, MPhil, PhD, is the Deputy Director of Research at Vinayaka Mission's Research Foundation (deemed to be university), India. Prior to this position, Dr. Suriyanarayanan served as Project Coordinator and Head of Department (I/c) at the Department of Water and Health, JSS Academy of Higher Education and Research, Mysuru, India. He also served as a science officer for SCOPE (Scientific Committee on the Problems of the Environment), Beijing Office, RCEES, Beijing, China, from April 2015 to March 2017. He is also serving as the nodal person for SCOPE India Activities. He has research experience in the areas of environmental monitoring, radiation ecology, and environmental microbiology. He secured the prestigious award of Visiting Scientist Fellowship of the Chinese Academy of Sciences during the years 2016–2017.

Contents

Preface

For the safe and responsible governance of our environment and its protection, there is an essential need for highly skilled labor in the field of environmental sciences. To develop the professional skills, there is a need to update existing knowledge and techniques, especially in how to accurately monitor environmental data and assess the change for the safe and responsible management of our environment.

This book is a collection of the latest research techniques on environmental monitoring and assessments. All submitted chapters were subjected to a peer-reviewing process. Chapters were chosen based on the quality and relevance of the themes.

I believe that the information contained in this book will enhance the skills of environmental scientists and decision makers and contribute to the exchange of best practices for developing and implementing optimum methods of analysis.

I would like to express my sincere appreciation to all the authors for their contributions to this book.

I believe that these chapters will be a good reference for academicians and researchers in the field of environmental sciences.

Special thanks goes to Dolores Kuzelj (Author Service Manager) for her dedicated support in the reviewing process and her suggestions for further improvement. Finally, all thanks to IntechOpen for publishing this book.

Suriyanarayanan Sarvajayakesavalu
Deputy Director Research
Vinayaka Mission's Research Foundation (Deemed to be University)
SCOPE (Scientific Committee on Problems of the Environment), India
Salem, Tamilnadu, India

Section 1

Water Quality Monitoring

Chapter 1

Hydrological Stress and Climate Change Impact in Arid Regions with Agricultural Valleys in Northern Mexico

Mariano Norzagaray Campos,
Patricia Muñoz-Sevilla and Maxime Le Bail

Additional information is available at the end of the chapter

http://dx.doi.org/10.5772/intechopen.80390

Abstract

In recent decades due to the negligence of groundwater management, the resources are degrading rapidly and have resulted in soil water stress. Owing to the role of climate change, now most of the environmental variables undergo variations in their dynamics and magnitude, and above all, arid zones suffer the greatest loss due to evapotranspiration and soil water stress. In order to understand the nutritional future of the most important agricultural regions in Mexico, this scientific contribution aims to analyze the trend of soil water stress in the Northern Mexico. The work emphasizes on the relationship between the energetic movement of groundwater and soil water stress which also includes the process of crop absorption that is affected by the changes in soil surface temperature and the geography of the region. The results reveal a positive trend that gradually leads to soil water stress and intensities associated with conditions of interdependence of unique variables in each valley. Global withdrawals of ground water tripled in the last 50 years with unequal volumes in arid zone that will definitely lead to instabilities, where sustainable management will be the basis of conservation to adapt to the new facies of climate change.

Keywords: groundwater, climate change, water stress, risk and agricultural valleys

1. Introduction

Worldwide, in many countries, the concerns of the productive sector on issues related to global warming and the consequences that current climate change could cause, especially the effects due to the changes in the average surface temperature of the world, will be discussed in this chapter. Soil and environmental temperature can significantly affect economic activities that are mainly related to the agricultural sector where soil is the main source of supply for food sustainability which is at risk due to water stress. The situation strongly affects the developing countries because of scarce economic resources that make their population vulnerable and poor resilience when it comes to facing the phenomena related to climate change. Decades ago, simple observations permitted mankind to realize the changes experienced by the environment, preferably by the increase in the frequency of occurrence of extreme weather events and environmental changes. Likewise, global environmental events associated with global warming caused the initiation of relevant investigations that would implement different methodologies to evaluate the vulnerability of climate change and henceforth to bring in awareness among the people. The book "Training methodology to assess vulnerability to climate change" is a typical example of this action. Undoubtedly, the frequency of occurrence and extreme meteorological phenomena of the past, present, and future and the fundamental cause is largely due to the increases and/or decreases in the regime of precipitation and potential evapotranspiration resulting in droughts and heavy rainfall regimes. Several research papers have documented the disparities and influences of climate change, whereas this work presents the thematic behavior of the temporal variation of soil water stress in the agricultural zones of North Mexico. In the present investigation, climate modifying factors in reduced spaces are analyzed in an integrated manner which could permit the scientific community to focus on general concepts such as heat islands and heat waves [1, 2] as in the past, which is actually related to augmented temperature rise [3].

There are many documented evidences regarding the economic losses caused by extreme meteorological events due to climate change, for example, Colombia witnessed 1970–2000 landslides and floods from 1970 to 2000 that caused heavy economic losses up to US $ 2.227 million that the National Planning Department (NPD) of that country was only 2.66% of the National Gross Domestic Product (GDP) for the year 2000 [4]. Perilous phenomenon in Mexico that could cause economic losses in the future, such as those caused in other regions of the country, is the excessive accumulation of cloud systems around the great mountains in the south of the country, preferably over the southern mountain ranges of Oaxaca and Chiapas, and also in the Northwest on the eastern and western Sierra Madre Oriental and Occidental. Several satellite images have also represented an increase in the spatial and temporal distribution of cloud systems (mostly clusters) that can reach up to size of a cyclone. The cloud accumulations before the presence of sufficient humid air initiate their route toward their distinct stages resulting in precipitation. However, irregular precipitations due to climate change can be considered to be the vital reasons that could increase the ambient and soil surface temperatures.

Specifically, in the mountain ranges of Sierra Madre Occidental and Oriental, the cloud clusters begin to group and tend to grow vertically to reach their mature state during the afternoons and later they begin to propagate throughout the region, but when crossing the great mountain ranges of the Sierra Madre, they form storms with intense lightning and up to the possible presence of hail. During the propagation, the clusters are conglomerated in large cloud systems of extensive oval shapes that can be observed by infrared satellite images. Likewise, the similar type of occurrences on August 23, 2010 caused heavy rains in the Papaloapan River Basin situated in the southeast part of Veracruz State, Mexico was observed, which eventually resulted in the accumulation of large oval cumulus systems during the same time period causing electrical storms that by the end of August ensued large floods in the Tlacopalcan community of the same state. Given the damage caused, there is no doubt about the importance of continuous monitoring of meteorological phenomena by means of available resources, and that the evolution and trajectory of the meteorological events must be tracked, in order to anticipate possible damage and take right decisions immediately. However, current meteorological studies do not reach such magnitudes of tracking, but with state-of-the-art technology developed in Mexico, satellite imageries and high-end monitoring techniques can provide early alerts and prevent risks from fires, severe storms, hurricanes, and also infrared satellite images can track the trajectory of a tropical storm in real time [5]. It is important to highlight that there are many scenarios in Mexico where the historical levels of diverse phenomena related to extreme meteorological events are surpassed; and most of them occur in regions which have experienced the rainiest periods in their histories leading to massive floods in the agricultural areas affecting the economy. Otherwise, the phenomenon can also occur in regions with absence of precipitation face drought scenarios such as those found in the northwest of the country facing Gulf of California. Regardless of the conditions caused by climate change, increase in the frequency of extreme meteorological activities always affects the economy and GDP of a country as climatic variability in great proportions distresses different productive sectors. Several studies at the regional level indicate that agriculture and rural development are severely affected, especially during the presence of the El Niño and La Niña phases; that increase or decrease in the climatic anomalies in large proportions bring floods, landslides in cultivated lands, proliferation of diverse pests, expansion of diseases, changes in the vegetative cycles of crops that are commonly enabled to the repeated ways of practicing agriculture, the seasonality of crops and production. Agriculture largely depends on the rainfall regime, soil surface and ambient temperature, and the soil conditions.

Therefore, it is logical to think that in places where climate change effects are observed, it can be inferred by the presence of environmental damages, so this would be the areas where the responses of soil due to climate change have to be defined, considering that soil, air, and temperature are in equilibrium. The large coastal plains of Northwest Mexico located along the Gulf of California encompassing the states of Sonora, Sinaloa, and Nayarit are the regions in which changes have been observed and reported for a long time due to the increased occurrences of extreme meteorological events. In the present work, the abovementioned regions are considered to be regions where geography intervenes to the formation of meteorological events. So, in this work, it is assumed that the environmental interventions due to climate change are directly proportional to the local geographical conditions that aid the formation of extreme meteorological events described above. That is the reason why we selected a coastal region of the state of Sinaloa to raise the objective of determining the change in water stress with respect to time [$\Delta Eh(x,t)$] in the soil of the valley of Guasave better known as "The agricultural heart of Mexico." The region is predominantly agricultural in which the subsoil has undergone agricultural pressure continuously since the 1960s due to the growing technologies and ideals set by the "Green Revolution."

In order to achieve the objective, data available from CONAGUA and the Environmental Engineering research group of CIIDIR-IPN-Sinaloa were analyzed and processed; in relation to the intrinsic variables of the agricultural land possibly affected by extreme meteorological events. The temporal space variation of $\Delta Eh(x, t)$ was measured based on the criteria of Medrano et al. [6] and conditioned to the methodological approaches of Sánchez [7]. The results can significantly measure over effects that could cause changes in the internal structure of the soil subjected to local climatic changes that could significantly affect the ways in which various crops exploit the H_2O resource for its development in the area possessing climatic conditions ranging from arid to semi/arid.

The investigation is justified by the mere fact of the latent presence of the meteorological phenomenon that currently occurs in both Sierra Madre Occidental and Oriental that causing rare meteorological events and forcing to carry out agriculture unceasingly under high risk.

2. Methodology

Among the three states of the Northwest Mexico (Sonora, Sinaloa, and Nayarit), a coastal agricultural region of Sinaloa State, Guasave, was selected for the present study. Located in the Northwest part of Sinaloa State, the region elucidates a long history of agricultural practices even from the period of "Green Revolution" (**Figure 1**). Due to its agricultural history, high productivity, and high-end technology, the area represents the incursion of large foreign currency into the Mexican economy, which have been the fundamental bases to be called "The agricultural heart of Mexico." The region consists of soils with diverse granulometric extents originated by the geological processes derived from different episodes that date from the Late Oligocene, the Middle Miocene (25–15 million years ago), and late Pliocene (<5 million years ago) and the current Quaternary [8]. The region also characterizes intermountain relief and coastal valleys.

The variation in the present-day relief started approximately in the altitudinal elevation of 1800 m, in the eastern limit of the Sierra Madre Occidental (natural barrier that protected Mexico against the dangerous tropical hurricane Patricia in mid-October 2015, formed by a tropical disturbance south of the Gulf of Tehuantepec). A transition towards the high plateau is characterized by a steep decline governed by an abrupt slope tending toward the Gulf of California and the Pacific Ocean until reaching the coasts at a regular altitude varying between 14 and 40 m above sea level. Specifically, to the North of the Sierra Madre Occidental, geological processes have resulted in an average altitude of 1400 m above mean sea level [9]. Henceforth, the area is made up of extensive agricultural coastal plains, and precisely the valley of Guasave constitutes slow and continuous flows of River Sinaloa streams of Ocoroni, Cabrera, and San Rafael; where the last three streams discharge into the Gulf of California. According to recent studies on the interpretation of geophysical data of geo-electric type [10], as a hydrogeological unit, the system is sometimes treated as a free aquifer that is commonly exploited and represents the regional geology which could be observed in varying depths where the hydraulic works are drilled [11].

Due to the changes in the environmental temperature, the increase and/or decrease of the precipitation regime, the presence of global warming and the consequences of climate change, some of the considerations of the criteria of Medrano et al. [6] were taken into account in order

http://dx.doi.org/10.5772/intechopen.80390

Figure 1. Coastal agricultural valley of Guasave, Sinaloa, Northwest Mexico.

to justify this investigation related to the ways in which a soil is stimulated $\Delta Eh\ (x, t)$. Given the presence and absence of the precipitation regime, it is considered that an environmental scenario can be altered under the behavior of the main factors that modify the climatic conditions in a reduced and closed space, and such a scenario was constructed in this investigation taking into account that the local geography is the fundamental cause that intervenes the modification of the main climatic factors. The important evidence that integrates these main climatic factors is to consider that it is a continuous system that fills everything within a closed space, that is, each of the variables measured at each point of the continuous system is the result of the volumetric response of the matter before the intensity of each variable that occurs within a closed space and continuous, in order to modify the factors that govern climate change and resulting in dissimilar soil conditions $\Delta Eh\ (x, t)$.

Since each continuous system fills every space it occupies, the magnitude of the variables that described $\Delta Eh\ (x, t)$ is the total response of effects on the matter or mass measured. Therefore, the accurate measurements of different variables that intervene in the modification of climate change and soil conditions were evaluated in the present study. The processes to characterize (physiological response of crops to water flow, capacity, and volume of each soil to store water and climatic conditions), were selected according to the criteria of Medrano et al. [6]

and various continuous variables were measured. Understanding the different physiological processes that govern water flow in plants allows us to estimate the efficiency of plants to manage water surplus and shortages. The density and depth of each plant species determines the water hold capacity of soil with respect to the total available water. However, the climatic conditions regulate the actions in which the plants need water and are strongly defined by the evaporation of the soil and in the same way by the atmospheric demand of how a soil requires water. The arguments by which the above criteria were selected was due to the fact that their variables can converge in an interdependent way and describe most of the processes that plants require during their development and growth, but also because they take into account a set of variables that intervene in the description of the physical state of the soil and which allows to act as indicators to estimate the behavior of $\Delta Eh\ (x, t)$.

The following is a brief description of the variables that intervene in the previous criteria and justifies the reason why they were taken into account in the definition of $\Delta Eh\ (x, t)$.

In the first criterion, the variables that intervene in the physical, chemical, and physiological processes inside and outside the plants are addressed and correspond to the variables that integrally give an ideal representation of the mechanisms that govern the water fluids. Therefore, it was considered that these variables define the availability of water resources, the humidity of both the soil and the environment. The variables considered were the depth at which the groundwater table is located in the aquifer as this distance is defined according to the internal and porous structure of the subsoil by 80% in the presence of the surface moisture of the soil and the moisture of the environment.

Regardless of the fact that a change in the precipitation regime is the one that can largely define the presence and/or absence of both surface and groundwater resources, this process does not result in a significant change in the internal structure of the subsoil at great depths and influences the physicochemical properties that would occur only in the structure of the superficial parts of the soil. Henceforth, within the aquifer, there are no significant changes in the ways in which advection processes can occur simply because no changes occur in the internal structure that defines the properties in which the advection occurs, to mention a few; storage coefficient, porosity, transitivity, hydraulic conductivity, among others; which will be discussed in the following criteria. Nevertheless, the foregoing, significant changes may occur due to the intensity and long periods of rain affecting the surface soils and within the aquifer, the changes would occur in relation to the magnitude of the intensity of the flows and quantities of existing fluids during the transportation processes in the porous medium. Therefore, it was taken into account that the magnitude of the following variables is those that contribute greatly to the existence of water flows and directly related to the temporal space variation of the depth of the static level of a region.

In the second criterion, the soil scenarios in which the crops are grown are also directly related with the internal structure of the soil and/or with the potential of each soil to exercise agriculture. Texture is one of the main variables that define the granulometric variation of each soil, degree of characterizing its structure type from thin to thick; that allow plants with the facility to extend their roots to the length of the subsurface horizons and are associated with the potentiality of crop development. Other variables that develop a crop into plant and aid in water storage are porosity (ϕ), hydraulic permeability (K), and transitivity (T) of the water resource through the porous medium that defines in a great way the opportunities of development of the crop and the amount of water required by the soil based on the availability of water, that could be altered by the presence of $\Delta Eh\ (x, t)$. Continuous water flow received by a plant is determined by T and ϕ of the subterranean environment, which according to Villanueva and Iglesias [12] T must be in a range of 500–1000 m^2/day. So, this research was based on the fact that the variables that define the development opportunities of a crop in terms of what is required to a high availability and/or absence of the water resource in the porous medium and in the plant would be the variation in the abovementioned parameters. Several field works were carried out in the area to define the variation of surface geology and to set up a monitoring network of 42 wells where the monthly sampling of water at different depths and soil for every 2 m was performed. Laboratory measurements were carried out to evaluate the porosity, transitivity, and hydraulic permeability in the saturated and unsaturated zone of the free and confined aquifer.

The third criterion was selected because it indirectly contemplates for its occurrence which includes the effects of the amount of radiation, humidity (absolute and relative), environmental temperature, and wind speed, which can be altered due to global warming, changes in the rainfall, and climate regime of a region. Regarding the modification produced by climate change, the criteria of Sánchez [7] were taken into account to select the factors that have enormous consequences on any climatic parameter and that depend on geography and that the literature considers them as the main regional climate modifiers. The modifying factors in a closed system that favor climate change were observed to be absolute height (*ah*), relative height (*rh*) or location, latitude (*Lat*), longitude (*Long*), exposure (*Ex*), and local continentality (*Ct*). The consideration by which we took into account these factors that can lead to climate change within a closed system was due to the fact that as a whole are the factors that depend on the local geography and can produce climatic mutations in a reduced space. Whereas in an open system, mutations are caused by the general circulation of the atmosphere (GCA) and they are generally very small and reduced to the degree which almost become nonexistent in such a short distance. Reduced climate change effects, unlike those that occur on a large scale, are favored by the conditions of the GCA, geographical factors, the shape of continents, and the action of the oceans. Therefore, the present investigation constitutes a basis to consider climate change as the sum of the set of microclimates influenced by closed spaces that function independently according to the local conditions and to the behavior of the main climate modifying factors already mentioned. Such climate modifiers are those that regulate the presence of water in a region, and according to the granulometric constitution of the soil, they would be those that as a whole affect the variations of the phreatic depth of the aquifer that regulates the presence or absence of environmental and soil humidity. Therefore, this work focused precisely on the exhaustive measurement of the phreatic depth in order to comprehensively describe the behavior of $\Delta Eh\ (x, t)$ considering other factors of vital importance of the soil, such as Tx and ϕ. The global climate change can be considered as a real representation of the climatic behavior at greater scales or by means of the summation or integral that makes a more complete estimate about the changes that occur globally. To understand this approach, it is important to establish that the main profound changes experienced by climate variables in a relatively closed area such as the Guasave Valley have a behavior that responds differently according to their geographical factors and directly influences the thermal and fluviometric conditions. In this study, $\Delta Eh\ (x, t)$ is considered as a relative system as absence and/or extreme excess of water in the soil can be considered already engrossed in the process of $\Delta Eh\ (x, t)$ and its behavior cannot be estimated through a simple statistical correlation or factor thresholds, but rather with the use of a complex statistical analysis, in which all the variables intervene and the interdependence and percentage contribution, with which the influence of each variable of the Medrano et al. [6] criteria can be described in an accurate way. The factors or variables are also influenced by the geographical effects according to a representative function (ψ (x,t)) of the integrated behavior of the system $\Delta Eh\ (x, t)$ in which the temporal space variation is the summation of the behavior of each variable as defined by the equation $Eh(x, t) = \oint_{to}^{tn} \psi\ (x, t)\ dx$, which integrates the changes occurred due to climate change in a closed system, where *to* and *tn* represent the change in time in the matter and that have effects on each variable to understand the intensive properties caused by climate change and as a whole can be considered as an extensive property as a function of time through each function ψ (x,t). The integral equation of the functions ψ (x, t) presents extensive properties in the matter that in this case is the soil and that in an integrated way the set of variables are those that define the response of $\Delta Eh\ (x, t)$ and the contribution of each of these is described mainly by the weight of each of the variables inside and outside the system. Furthermore, if the property of the material can be extensive in time, if and only if it is governed by the behavior of the previous equation, it can be said that the effects caused by climate change respond to two properties in the matter that is associated to "the intensive property given by the variables and the extensive property given by the effects of climate change." In this case, according to the principle of continuity, the intensive properties are the variables resulting from continuous interaction of climate change with the matter resulting in different conditions of $\Delta Eh\ (x, t)$, whereas extensive properties refer only to the effects that take place in the set of all the variables before the intrinsic properties of the soil and they result in the total presence of $\Delta Eh\ (x, t)$. Thus, the relationship presented is an integral function that defines an intensive property and establishes a one-to-one correspondence between extensive and intensive properties. In particular, if the values of the integral function are vector, then the corresponding extensive function will also be vector. It should be noted that there are different ways to define an intensive property and, in our case, we have defined it as the property that has the effects of climate change per unit volume.

From the physical point of view, the basic hypothesis to formulate this investigation was "The balance of the extensive properties in the theory of the systems of the continuous flows and of the variables that interact with the matter, as a whole converge to define an integrated manner both in closed and open systems the effects produced by climate change" whose numerical analysis will be established through the type of system (closed and/or open) and its mass balance can be obtained through the premise "any variation of the extensive property caused by the changes in matter due to climate change is the result of what is generated or destroyed within the body or mass and of what enters or leaves through its borders." Given that the effects caused by climate change depend on the main geographic factors that regulate the climate of each region, precipitation, and environmental temperature, their respective magnitudes were concentrated in a database to position the variables according to the three criteria of Medrano et al. [6] and according to the territorial extension. To perform the measurements of the variables involved in the description of $\Delta Eh\ (x,\ t)$, a monitoring network of 660 wells was established. Because of the costly, laborious, and/or simple that it was to obtain the sample and how difficult and/or easy to access, from the general network in a random way, 42 wells were selected, which were taken monthly at every 2 m of depth. Water and soil samples were transferred to the laboratory to measure the physical and granulometric characteristics, respectively. Following the techniques of Bouyocus and paraffin, the values of Tx and ϕ, respectively, were determined. In the network of 42 wells using the simplified Theis method, pumping tests were carried out to estimate the magnitudes of T, S and K, while in the network of 660 wells, we used an electrical probe to measure the phreatic depth and at the same time the height of the curb of each hydraulic work was also determined. With the measurements of phreatic depth and height of the curb, the hydraulic load (H) was obtained in each well and using Darcy's Law, the regional hydraulic gradient (∇.H) was obtained from well to well; and thus, through the interpolation of data by means of a Kriging, maps were prepared that indicate the preferential direction of the groundwater flow from the recharge zones to the discharge zones.

During the field trips, the locations of the wells in both the networks were estimated using differential GPS. The temporal variation of the solar radiation was also measured using a pyrometer. Although a good number of variables were measured to understand the dynamics of $\Delta Eh\ (x,\ t)$, the statistical analysis was focused on the description of the behavior of the phreatic depth of aquifers, since it was considered as one the most important variable responsible for the volumetric exchanges of H_2O between the aquifer and the surface; besides the absence and/or presence of this in the porous structure of the soil, in the plants as well as in the environment. Therefore, the proximity or phreatic distance was considered directly proportional to the moisture potential of plants. To describe the behavior of the zone of saturation, statistical measurements of central tendency, noncentral position, and dispersion were used [13]. The measures of central tendency used were Mean ($\overline{X}$), Median (M_e), Kurtosis (K_s), and the dispersion of range of variation ($X_{min,}\ X_{max}$), Standard deviation (σ_0), and Variance (σ^2). The statistical measures carried out using the software STATISTICA (version 7.0) and the interpolations to obtain the equipotential curves of the variables were done using SURFER (version 10.0), where the latter formed to be the fundamental basis for the preparation of the final maps attached to the regional urban trace. Final details were elaborated using the program Corel Draw X6.

3. Results and discussion

To meet the first criterion of Medrano et al. [6] for the presence and/or absence of surface humidity in the soil and in the environment, indistinctly of precipitation, the results of the spatial variation of the phreatic depth are discussed. The phreatic depth variation from the measures of central tendency presented an irregular spatial behavior along the valley with respect to the average value obtained from the dataset in the network constituted by 660 wells from the coast to the mountain areas. The central tendency measurements marked somewhat similar values for $\overline{X}$ and M_e equivalent to 5.88 and 5.61 m; measures of central tendency did not provide an adequate description of the set of measurements that defined the spatial distribution of phreatic depth in the Valley, and could not give an exact description of the variation of this parameter; however, because the phreatic depth values turned out to be superficial, it can be regarded as a potential indicator of the evidence of $Eh(x, t)$ type saturation in most of the soils of the study area. The values of $X_{min,}$ X_{max} were 0.94 and 16.17 m, respectively, whereas the variability or dispersion of the variance was observed to be $\sigma^2 = 10.34$ m and $\sigma_0 = 3.21$ m with $E\sigma_0 = 0.03$, which was relatively low.

The values of σ_0 were accepted to be less than the values for $\overline{\overline{X}}$ and M_e, situation observed by the symmetry of the phreatic distribution along the valley with the magnitude of 2.54 presented the distribution of Ks. σ_0 showed a less value, so in general terms, the data collected were considered adequate for the information process. The spatial variation of the phreatic depth is presented in **Figure 3** and the lesser values of the phreatic depth ranged from 1 to 5 m with parallel alignments to the coast line and in the preferential direction of NW-SE revealing the protuberances of the Sierra Madre Occidental. These conditions of low depth represent the presence of a high environmental humidity; however, it is also indicative of high saturation states in the internal structure of the soil and in turn the presence of $\Delta Eh(x, t)$, in which the soils by proximity to the sea in the presence of an intrusion phenomenon are no longer apt for practicing agriculture. In turn, the spatial configuration of the phreatic zone allows us to observe soils in transition to the presence of $\Delta Eh(x, t)$ and that could be considered as a future risk. Nonetheless, the foregoing, before the most important factors that modify the climatic conditions of a closed environment, it is important to take into account the superficial granulometric variation of the soil that may constitute its full conservation or its ease of destruction, and which is defined respectively by the magnitudes of Tx and ϕ; also indicate the potential of the soil to retain H_2O, so they are both directly related to the field capacity of the soil that is an indicator of the amount of water moisture that the soil requires after being completely saturated, may be it has been wetted and drained with respect to the topographic slope until again the water potential stabilizes its action that lasts from 1 to 2 days (24–48 h).

There are evidences that at the depth at which the phreatic zone is located, a very specific humidity condition will be defined in the environment that is directly proportional to its depth, i.e., there is a linear and direct relationship with soil moisture. Due to the above conditions, increases and decreases in the water table of a zone like those observed in **Figure 2** can represent a direct relation to the period of time without precipitation, where Tx and ϕ play an important role in the filtration/runoff interaction, being greater and faster the response the shorter is the period. Similarly, considering the logic to a hydrological response of the soil of a valley in the presence of rainfall, it is expected that the phreatic zone will vary in large scale depending on the time in which the rain occurs with respect to the previous period, and this hydrological response of the soil has increased to the extent that there are increases in phreatic zone, or decreases before the decrease; therefore, the speed of the believed hydrological response will depend on the forms of how Tx and ϕ permit it.

In **Figure 3(A)**, the spatial distribution 660 wells where the measurements of the depth of the water table (*m*), hydraulic conductivity (*K*, m/s), transmissivity (*T*, m^2/s), curb height (*hb*, msnm), hydraulic load (*H*, dimensionless), and hydraulic gradient (∇.H) are presented. The

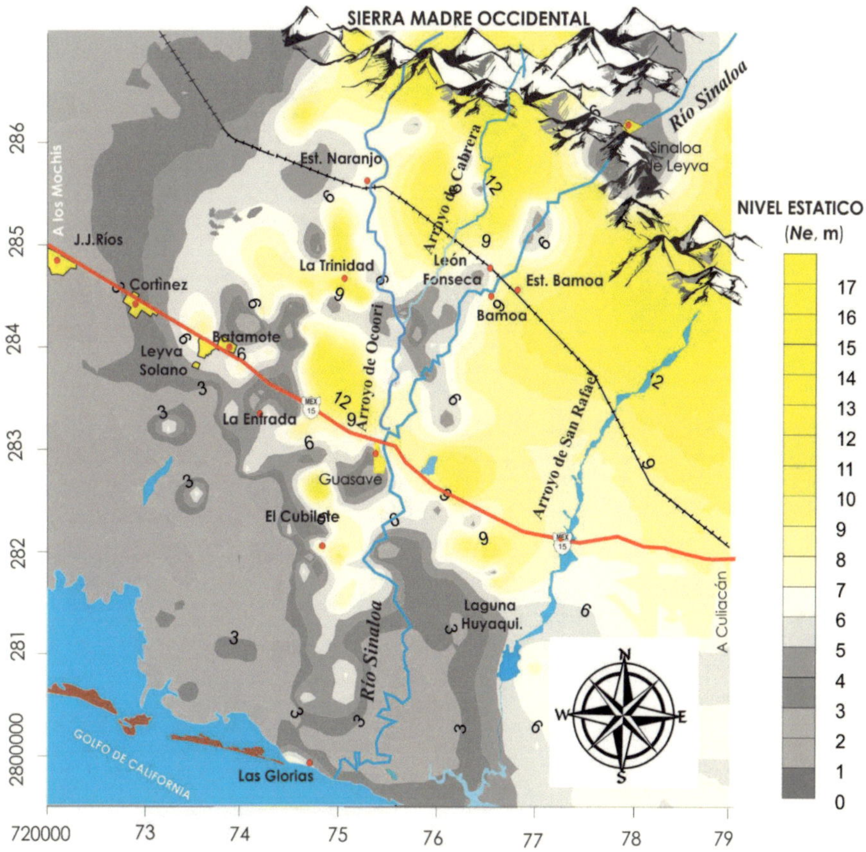

Figure 2. Spatial variation of the depth of the static level (*Phreatic depth*, m), configured for 2018 using 660 available wells for both human consumption and agricultural use (wells and deep agricultural wells) from the coastal zone of the Sierra Madre Occidental.

last three parameters were important in the determination of the piezometric map indicating the preferential direction of the waters in the subsoil. In part B, of the same **Figure 3**, the monitoring network of 42 wells selected from the general network is shown, under the consideration that there is an ease in the road and gaps to achieve access to the wells and facilitate the transfer of samples to the laboratory to perform the determination of the texture parameters (*Tx*, dimensionless) and porosity (ϕ, %) of the soil. Parts C and D of **Figure 3** present the map of the spatial distribution of the parameters *Tx* and ϕ measured in the laboratory. The last two maps are the result of the "Kriging" type interpolation made in the SURFER 10.0 computer program.

The reason why a wide network of 660 wells was selected was due to the fact that it was considered to have a greater control in the phreatic depth of the aquifer due to the fact that most of the water is concentrated throughout the year, and it is with respect to the phreatic depth. This network constitutes a constant and important source of water volumes that intervene in the content of the environmental and soil humidity, as well as large volumes of water for the ETP processes occurring between the plant and the soil.

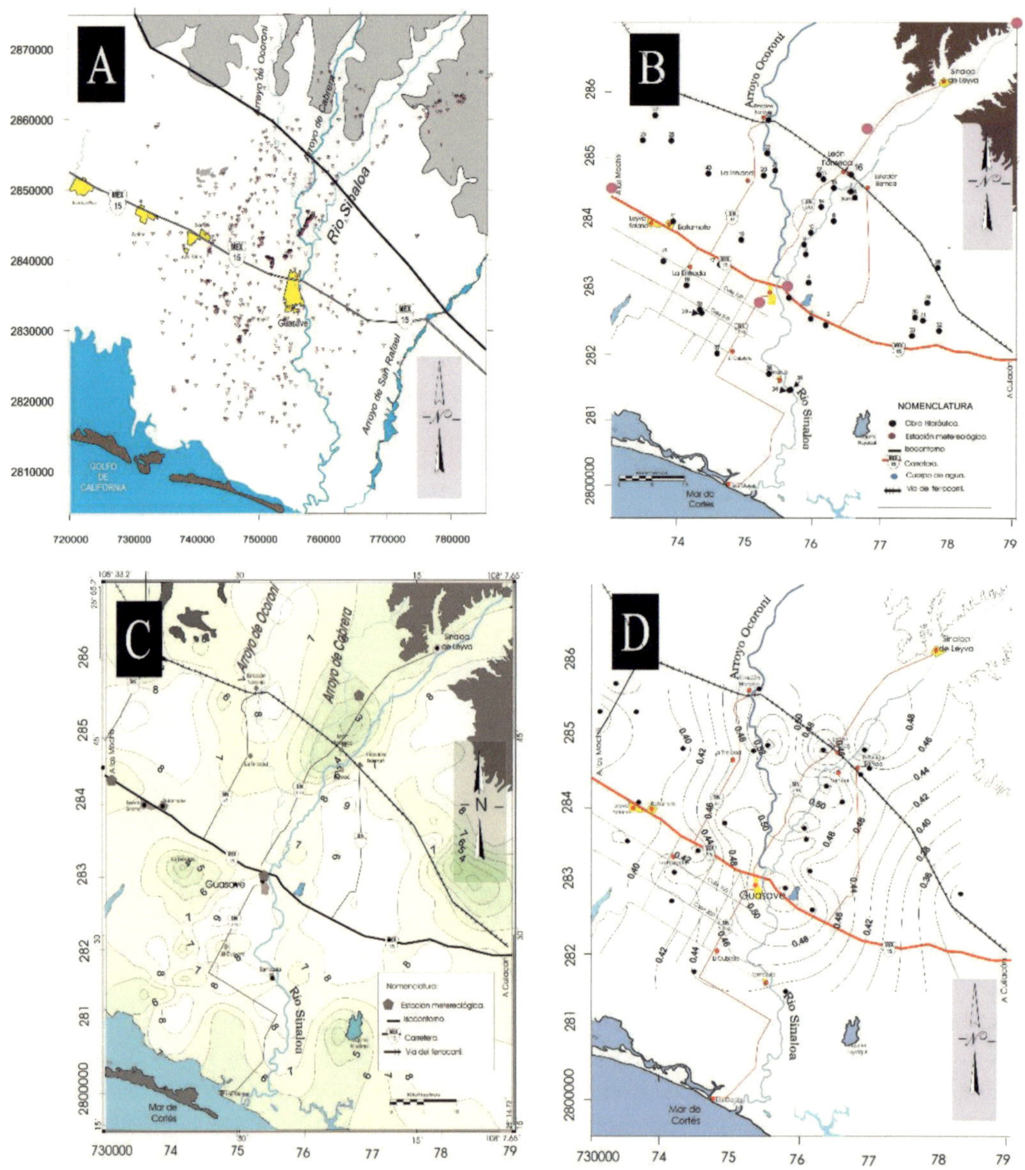

Figure 3. (A) General monitoring network of 660 wells selected randomly for the study (B) 42 wells selected for the measurements of textural parameters and porosity of the soil; (C and D) show the behavior of the spatial variation through the equipotential curves of textural parameters and porosity.

Figure 3(C) and **(D)** represents the spatial behavior of the equipotential curves for the values of *Tx* and *ϕ*, both obtained in the laboratory with their respective techniques. Since most of the crops in the valley are concentrated in the central areas of the valley, in the search of the best representation for *Tx* and *ϕ*, it was decided to have the majority of the wells distributed in random form in the central area as presented in **Figure 3(B)**. **Figure 3** presents the spatial variation of *Tx* for which a scale was established, ranging from 1 to 11, which allowed to analyze the granulometric variation that goes from the most compact, through the soils of intermediate type until reaching the lowest content of cement: (1) clay, (2) sandy clay, (3) silty clay, (4) clayey-sandy mudstone, (5) clayey mudstone, (6) clayey-silt mudstone, (7) sandy mudstone, (8) mudstone, (9) silty mudstone, (10) silt, and (11) sand. If it is taken into account that the soils have the highest concentration of cement between their grains, they are soils that have a tendency toward low values of K, T, and S, so this type of soil will be revealed in the drought seasons.

According to the values of φ (**Figure 3(D)**), a heterogeneity will be exhibited due to the high values of K allowing the soil to reach scenarios of $\Delta Eh(x, t)$ of high water saturation or low values of φ with low K values that tend to be completely dry. Therefore, correlations between intensive and expansive processes in the gradual scenarios of $\Delta Eh(x, t)$ can be represented through the following multiple and interdependent correlations $[tx\ vs. K\ vs. \Phi\ vs. T\ vs\ S]$. It is also important to mention that the porous medium and its changes with respect to its internal structure are made in a slow way and are minimal changes that may occur over long periods of time; however, in the presence of extreme meteorological events, soils present erratic portrayals.

Despite the slowness with which the internal structure of the soil occurs, a scenario of drought or saturation of its internal structure is highly conditioned to the availability of water resources, and the aforementioned multiple interdependent correlation would be dominated exclusively by a scenario of drought or saturation, with all the intrinsic parameters of the soil to be defined $\Delta Eh(x, t)$ under the availability of water resources. In the map of Tx, in **Figure 2(C)**, an alignment in the direction NE-SW parallel to the flow of river sediments with the tendency of belonging to sandy soils can be observed, also the similar scenario is repeated in the area located to the NW and in some areas of the coastal zone. The clayey soils are observed to be distributed concentrically along the valley, which indicates isolated basin lines where the water is concentrated to drain quickly toward the tributaries of the area, not allowing the soils a strong interaction with the water resource which also puts them at risk of being dry. The region is partially covered by alluvial materials and fluvial deposits of the Quaternary, which occupy the subsoil of the entire region with variable thicknesses, heterogeneous in terms of lithology, degree of cementation, and hydraulic characteristics.

Figure 4 represents the water availability (mm^3) in the valley. The statistical measurements in the distribution for the central tendency of the data describing the availability of H_2O showed values for x^- = 1337.92 mm^3 and M_e = 1095.91 mm^3, with very significant variations among each other; so, this type of measures of central tendency revealed the first approximation that the values of the distribution do not meet with a dominant or preferential tendency toward the minimum or maximum values of the whole distribution and propitious to reach in a gradual way the different scenarios of $\Delta Eh(x, t)$. Water availability measurements are an indicator of the existence of a marked variation in the availability of the H_2O resource within the aquifer, which controls the presence of moisture in the soil and environment. The observations were complemented by the statistical dispersion and noncentral measurements, X_{min} = 109.12 mm^3; X_{max} = 3194.33 mm^3; σ^2 = 599082.97 mm^3; σ_0 = 774 mm^3 (defined by $E\sigma_0$ = 8.2), and Ks = 2.5. The manifestation of σ^2 and high σ_0 above 50% in the values of x^- and M_e define a different availability of the water resource from one place to another, in such a way that the soil can have different values of absolute and relative humidity in the states of $\Delta Eh(x, t)$.

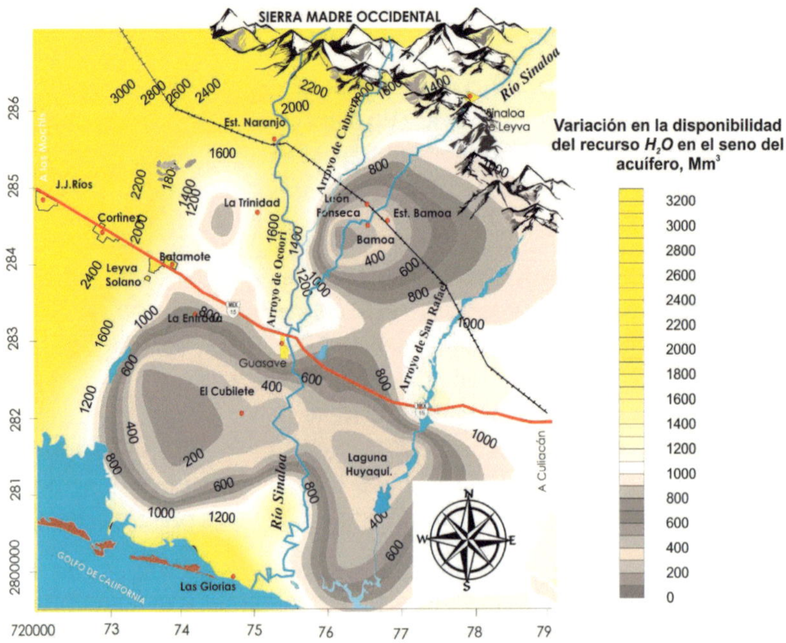

Figure 4. Spatial variation of the available water resource in the aquifer of Sinaloa river valley.

Figure 4 presents the North and Northwest part of recharge of the area, as well as the discharge area that is mostly located toward the most important water tributaries; the recharge was observed to be 1600 mm^3 and reaches a volume of up to 3200 mm^3.

The marked variation in the water availability of the central area of the valley is most probably to the sedimentary terrain of the region. High volumetric imbalance in the storage coefficient can be related to the high demand for groundwater that has always existed in the region due to extensive agriculture practices, livestock, and trade business. It should be noted that these soils, due to their varied granulometric composition, are the soils that are preferred in agricultural activities because they allow diversity in the types of crops of commercial interest whose productivity provides guarantees of having diverse products throughout the year.

Differences in the agricultural activity governed by different types of soils throughout history have been favored by a superficial sedimentary geology in which the water tributaries play an important role in the formation of its granulometric efficiency. In the present study, soils are mostly characterized by a silty-sandy mudstone and silty-clay mudstone, making them suitable for agricultural activities as they permit water content in the pores that have no tendencies toward positive values and extreme negatives of being able to store H_2O in its structure and are suitable for any type of crop growth. The variations that exist in the availability are relatively of small gradient (approximately 200–300 mm^3 for every 2–3 km), defining the area as high risk tending to the presence of $\Delta Eh(x, t)$ primarily subjected to water availability; however, in a previous year, a scenario with high intensities of precipitation had been presented.

4. Conclusion

The current increase in temperature and changes in weather conditions had significant impact on the rainfall. The absence/increase of rainfall has possibly been resulted in flooding and/or drought. The soil in these regions has faced different "stress states" depending on the rainfall. So, it is always important to know its current response to the new precipitation regimes, in order to identify the porous internal structure and the current potential to store the water resources. Usually, the development of crops is based on the presence of maximum or minimum H_2O retention scenarios and/or absence of the relative and absolute humidity. One of the methods by which they form extreme meteorological events as mentioned in the introductory part, it is directly associated with the formation of cumulus clouds. Occurrence of meteorological events increases when the clouds reach mature stage that causes high cumulus densities to accumulate in the atmosphere by setting the surrounding areas at risk with the continuous manifestation of high rainfall and droughts. It occurs mainly in the presence of humidity in the environment. Continuous accumulation phenomena of cumulus clouds and the saturation of the ground that caused increases in the conditions of $\Delta Eh(x, t)$ depend on the soil moisture condition, the behavior of the water table, surface and subsurface processes of the soil, and the environment. Thus, it is important to know the conditions in which it can develop the magnitude of $\Delta Eh(x, t)$ and are produced to a certain extent by the conditions of antrum factors depending on the activities of each region. The soil protection programs would preferably be aimed at protecting its internal structure through nondeforestation of green areas. Another way is to avoid the production of gases favoring the greenhouse effect, which in turn attributes further changes in global temperature. Understanding the given essentiality of soil in the world, humans must adapt and adjust with respect to climatic stimuli in order to moderate the damages happening in the environment. In addition, it is worthy to identify the beneficial opportunities and possible resources to circumvent the new changes. On the other hand, the current flora and fauna is the result of a continuous evolution by an adaptation to the different environmental conditions, i.e., changes in the temperature on the face of the Earth. We suspect that these new meteorological processes like the one presented here would lead to environmental changes affecting plants. As plants are mostly in contact with soil, water, and air, they are easily amenable for adaptations to environmental temperatures and the availability of the water resources. The governance in tolerance and the way of developing adaptability will allow the plants to define their permanence or disappearance as a species. Apparently, in a given land, fundamental factors like, the dynamics of changes undergone by a land through its history in the different stages of its evolution, and in some of the occasions, external conditions from other planet have been considered to identify the sources causing changes. On this occasion, the evolution dynamics can be conditioned to the high and/or low availability of the water resource and to the environmental conditions in which the soil develops the phenomenon $\Delta Eh(x, t)$. Currently, many scenarios are changing in relation to the ambient temperature which has forced the plants to develop new tolerance limits and live in the presence or absence of H_2O.

An increase or decrease in the ambient temperature, in different regions, will allow the existence of new living conditions experienced by the scenarios that will be conditioned by the presence and/or absence of the superficial or underground H_2O depending on the phreatic depth and the water retention capacity of different types of soil. Henceforth, it is recommended in the places where the ambient temperate decreases or increases, studying the availability of water resources from a point of view where the variables are related by the geographical effects allow the development of climatic conditions that are different for both global and closed environment, so the representative function of the change $\psi\ (x, t)$ according to each variable will tend to behave differently from the effects caused by the geographical factors but there will always be a function in each environment (representative of the integrated behavior of the system $\Delta Eh(x, t)$, in which the general spatial temporal variation of the global climate change can be represented by the summation of the behavior of each variable defined in the $\Delta Eh(x, t) = \oint_{to}^{n} \psi\ (x, t)\, dx$). It is concluded based on the argument of the equation that integrates $\psi\ (x, t)$ as a whole through its environmental phenomena having energetic properties, so they can produce changes in the extensive properties of matter, which in this case is the soil; and one of those changes in the set of variables that define the response of the soil to $Eh(x, t)$. The energetic process attributed to the changes in matter is found or conditioned mainly by the weight that each variable has inside and outside the soil system subjected to the continuity principle of the intensive property that interacts and formulates climate change with effects before the subject leading to different status of $\Delta Eh(x, t)$ referred exclusively to the result caused by the extensive properties of matter between the effects produced by all the variables in relation to the intrinsic properties of the soil, which would be a response to anisotropy, heterogeneity, and interdependence on the forms in which $\Delta Eh(x, t)$ appears in the different soil types. Therefore, given the moisture content of the soil surface and the environment, since both are conditioned to the phreatic depth, a change in land use may affect the magnitude and seasonality of the storm flows and the seasonality of the plants and crops, making them difficult to evaluate due to the high variability in the response of the watersheds to the events of precipitation and the state of soil moisture, especially when evaluating with respect to the previous event, as Bruijnzeel [14] attributes that the observations in the variations of the humanity have observed in the different places of the world that have basins experiencing extreme precipitation events. Real quantitative predictions cannot be made for any area particularly since many of these subjected to climate variability [15, 16]. Therefore, there is a need to implement specific monitoring programs for hydrological variables focused on the identification and quantification of processes, in order to determine the effects on the flow dynamics in relation to the coverage and conditions of the basin or its state of alteration. So the availability of water resources in different parts of the earth is defined through the hydrological cycle conditions of H_2O, and the magnitudes of different precipitation and evapotranspiration vectors that, together with other factors, will define the ideal amount retained in the subsoil thus favoring the development and adaptation of the flora and fauna that characterizes each place. The decrease in water availability results in aridity and when the pluviometric manifestation occurs in large volumes that exceed the saturation threshold of the soil, flooding occurs. At present, concepts of drought and a weather condition in the absence of rain can be tolerated by all plants. Water deficit refers to the water content of a tissue or cell in a plant below the highest water content that can resist, and is exhibited in the state of greatest hydration [17]. The complexity in the analysis of this type of phenomena is attributed to the forms that originate them, as well as the lack of detailed quantitative information on the changes in the infiltration capacity of the different soil types and the fertile profiles of hydraulic conductivity with respect to the soil

http://dx.doi.org/10.5772/intechopen.80390

depth, moisture retention capacity, and the depth of development of plant roots, in front of climate change risks, the presence of systematic sampling campaigns is demanded in order to perform numerical models of associated extreme flow rates in hydrological basins.

Author details

Mariano Norzagaray Campos[1], Patricia Muñoz-Sevilla[2]* and Maxime Le Bail[2]

*Address all correspondence to: npmsevilla@gmail.com

1 Centro Interdisciplinario de Investigación para el Desarrollo Integral Regional (CIIDIR), Unidad Sinaloa, Instituto Politécnico Nacional (IPN), Guasave, Sinaloa, México

2 Centro Interdisciplinario de Investigaciones y Estudios sobre Medio Ambiente y Desarrollo (CIIEMAD), Instituto Politécnico Nacional (IPN), Ciudad de México, México

References

[1] Jones PD, Mann ME. Climate over past millennia. Reviews of Geophysics. 2004;**42** (RG2002):42. DOI: 10.1029/2003RG000143

[2] Stott PA, Stone DA, Allen MR. Human contribution to the European heat wave of 2003. Nature. 2004;**432**(7017):610-614

[3] International Food Policy Research Institute (IFPRI). Climate Change Impact on Agriculture and Costs of Adaptation. Washington, D.C.; 2009. p. 30. Available at: http://www.fao.org/fileadmin/user_upload/rome2007/docs/Impact_on_Agriculture_and_Costs_of_Adaptation.pdf 6

[4] Cárdenas M, Rodríguez M, editors. Desarrollo Económico y adaptación al cambio climático. Bogotá, Colombia: Frescol and Foro Nacional Ambiental; 2013. Available at: http://www20.iadb.org/intal/catalogo/PE/2013/11695.pdf

[5] Prieto R, Oropeza RP. Estudio para el desarrollo de nuevas herramientas para el Sistema de Alerta Temprana de Ciclones Tropicales. México: CENAPRED, Subcoordinación de Hodrometeorología, Coordinación de Hidrología; 2007. p. 42. Available at: http://www.cenapred.unam.mx:8080/TransparenciaGobMX/documentos/Renglon_10.pdf

[6] Medrano H, Bota J, Cifre J, Flexas J, Ribas-Carbó M, Gulías J. Eficiencia en el uso del agua por las plantas. Investigaciones Geográficas (Esp). 2007;**43**(1):63-84. Available at: http://www.redalyc.org/pdf/176/17604304.pdf00

[7] Sánchez MJ. La creación de sistemas climáticos actuales en función de factores geográficos propuesta metodológica. Lurralde: Investigación y espacio. 1994;**17**(1):229-241. Available at: http://www.ingeba.org/lurralde/lurranet/lur17/17sanche/17sanch.htm

[8] Ferrari L, Valencia-Moreno M, Bryan S. Magmatismo y tectónica en la Sierra Madre Occidental y su relación con la evolución de la margen occidental de Norteamérica. Boletín de la Sociedad Geológica Mexicana. 2005;**57**(3):343-337. Available at: http://www.geociencias.unam.mx/~alaniz/SGM/Centenario/57-3/(5)Ferrari.pdf

[9] González-Elizondo MS, González-Elizondo M, Tena-Flores JA, Ruacho-González L, López-Enríquez IL. Vegetación de la Sierra Madre Occidental, México: una síntesis. Acta botánica mexicana. 2012;**100**:351-403. Available at: http://www.scielo.org.mx/scielo.php?script=sci_arttext&pid=S0187715120120003000012&lng=es&tlng=es

[10] Herrera-Barrientos J, Mendoza C. Determinación de la conductividad eléctrica del subsuelo mediante inducción electromagnética en la zona cercana. Informe Técnico, Comunicaciones Académicas, Serie Geofísica Aplicada. CICESE; 1995. p. 55. CTGAT9503

[11] Norzagaray CM, García GC, Muñoz SP. Impacto natural-antropogénico en el flujo y los niveles piezométricos del acuífero del río Sinaloa. Revista Latinoamericana de Recursos Naturales. 2009;**5**(3):212-218. Available at: http://itson.mx/publicaciones/rlrn/Documents/v5-n3-3-impacto-natural-antropogenico-en-el-flujo-y-los-niveles-piezometricos-del-acuifero-del-rio-sinaloa.pdf

[12] Villanueva MM, Iglesias LA. Pozos y acuíferos técnicas de evaluación mediante ensayos de bombeo, Talleres gráficos IBERGESA. 1984. p. 426. Available at: http://aguas.igme.es/igme/publica/libro35/lib35.htm

[13] Begoña BF. Análisis de los datos en un proyecto de investigación. Matronas Profesión. 2005;**6**(3):30-36. Available at: http://www.federacion-matronas.org/revista/matronas-profesion/sumarios/i/7069/173/analisis-de-los-datos-en-un-proyecto-de-investigacion

[14] Bruijnzeel LA. Hydrological functions of tropical forests: Not seeing the soil for the trees? Agriculture, Ecosystems and Environment. 2004;**104**(1):185-228. DOI: 10.1016/j.agee.2004.01.015

[15] Elkaduwa WKB, Sakthivadivel R. Use of Historical Data as a Decision Support Tool in Watershed Management: A Case Study of the Upper Nilgawa Basin in Sri Lanka. IWMI Research Report No. 2 (Vol. 26). Colombo, Sri Lanka: International Water Management Institute; 1999. p. 31. Available at: https://books.google.com.mx/books?hl=es&lr=&id=Wnp43jePLAMC&oi=fnd&pg=PR5&dq=Elkaduwa,+W.K.B.+y+Sakthivadivel,+R.+1999

[16] Richey JE, Nobre C, Deser C. Amazon river discharge and climate variability: 1903-1985. Science. 1989;**246**:101-103. DOI: 10.1126/science.246.4926.101

[17] Taiz L, Zeiger E. Plant Physiology. 4th ed. Sunderland, Massachusetts: Sinauer; 2009. p. 719

Chapter 2

Evaluation of Water Quality Indices: Use, Evolution and Future Perspectives

Carlos Alexandre Borges Garcia, Igor Santos Silva,
Maria Caroline Silva Mendonça and
Helenice Leite Garcia

Additional information is available at the end of the chapter

http://dx.doi.org/10.5772/intechopen.79408

Abstract

The evaluation of the quality of water bodies is of fundamental importance to the study and use of water. Aiming to improve the understanding of the phenomena which occur in these environments, several indices have been proposed over the years, using several statistical, mathematical and computational techniques. For this, it is necessary to know the variables which influence different water bodies. However, not all places are able to make the most diverse analyses due to the financial and sanitary conditions, which can promote greater expenses in treatment as well as make the limits of tolerance of the water quality higher. Nowadays, there is a need to formulate indices which can address climate change in its variables, making it even closer to reality. In this context, seeking to reduce the number of variables used, collection costs, laboratory analyses and a greater representativeness of the indices, multivariate statistical techniques and artificial intelligence are being increasingly used and obtaining expressive results. These advances contribute to the improvement of water quality indices, thus seeking to obtain one which portrays the various phenomena which occur in water bodies in a more rapid and coherent way with the reality and social context of water resources.

Keywords: water quality index, statistical techniques, machine learning, artificial intelligence, environmental monitoring

1. Introduction

Several quality indices are proposed for evaluation and definition of different uses of water. These proposals are interesting so that the selection criteria of the parameters are more effective and portray the true environmental state of the water body. In addition, other important factors in this selection are the availability of analysis of variables by laboratories (if they have equipment, reagents and staff for such analysis), the financial question of the region (poorer regions have more flexible tolerance limits, whereas that water treatment is not as efficient), collection logistics and representativeness for an audience which often does not have the ability to interpret the analysis variables, thus requiring the use of quality indices.

These indices ought to be elaborated quickly, simply and objectively. In this way, the use of water quality indices (WQI) is fundamental to represent a large number of parameters in a single number. However, for this number to portray the reality of a water body, the correct selection of environmental parameters is essential.

This selection of environmental parameters ought to be fundamental not only for the elaboration of indices, but also for the monitoring of water resources. This monitoring has been understood as the definition of strategies to mitigate environmental problems which can guarantee sustainable development. A water quality index inserted in the context of monitoring also makes it easier to reproduce the information for those involved in both management and those who make direct use of water.

In this context, the modeling and construction of a single index which frames the water bodies, whether these are lentic or lotic, involve several factors which combined make difficult the existence of this uniqueness. The morphoclimatic diversity of each region of the world, climatic changes and innumerable anthropogenic activities, as well as local social and economic development are preponderant factors for different classifications of water quality and, consequently, several indices. In order to face these adversities, multivariate statistical techniques and artificial intelligence play a fundamental role, facilitating the framing and understanding in the search for a globally accepted index.

In this sense, in order to reduce the number of environmental parameters to be measured, the costs involved in collecting the water samples and the laboratory analyzes, in addition to seeking greater representativity of the indexes, multivariate statistical techniques and artificial intelligence have been used and the results have been significant and promising. Such techniques allow greater agreement between the variables most suitable for the formulation of an index and ought to be part of an environmental monitoring program.

2. Methods

2.1. Method of selection of variables

The water quality index (WQI) was initially proposed by Horton [1] as a linear summation function. This index consisted of a weighted sum of subscripts, divided by the sum of the weights multiplied by two coefficients related to temperature and the pollution of a watercourse. Horton [1], in his work, used as criteria of choice the variables most used in the analysis of a water body in a total of 10, making the application of the index more practical. These variables should represent all the water bodies in the country and should reflect the availability of the data, in order to obtain the smallest deviation among them [1–5].

In order to construct an index, four steps are taken: parameter selection, obtaining subindices; establishment of weights; use of aggregation functions. A criteria was developed based on the existing indices, as presented in **Table 1**. Although indices are not expected to meet all these criteria, different weights can be attributed to each one, depending on the use, region, climate and water body. Thus, with the definition of weights, the criteria can be used to elaborate the index as comprehensively as possible [4, 6, 7].

2.2. Main indices

Based on the index proposed by Horton [1], Brown et al. [8] proposed the best-known and most widely used index in the world, i.e. the National Sanitation Foundation's Water Quality Index (WQI-NSF). This index can be used to define water quality for different uses such as irrigation,

Number	Criteria
1	Relative ease of application
2	Balance between necessary technical complexity and simplicity
3	Present an understanding of the significance of the presented data
4	Include widely used and routinely measured variables
5	Include variables that have clear effects on aquatic life, recreational use, or both
6	Inclusion of toxic variables
7	Easy introduction of new variables
8	Based on recommended limits and water quality standards
9	Developed with a logical scientific reasoning or procedures
10	Tested in a number of geographic areas
11	Present agreement with expert opinion
12	Demonstrate compliance with biological water quality measures
13	Dimensionless
14	Well-defined intervals
15	Possessing statistical property which allows probabilistic interpretations
16	Avoid the eclipsing effect
17	Present sensitivity to small changes in water quality
18	Present tendencies over time aiming for applicability for comparisons of different locations and for communication with the public
19	To present ways of dealing with the absence of data
20	Clearly explain the limitations of the index

Table 1. Criteria established for the elaboration of an index [4, 6].

Parameters	Weight
DO	0.17
Fecal coliforms	0.15
pH	0.12
BOD	0.10
Nitrate	0.10
Total phosphate	0.10
Temperature	0.10
Turbidity	0.08
Total solids	0.08

Table 2. Parameter weights as index [20].

water supply and navigation, as well as for various water bodies (lakes, reservoirs and rivers). In this index, nine parameters were used according to the criteria presented in **Table 1**: temperature, pH, turbidity, phosphate, nitrate, total solids, dissolved oxygen (OD), biochemical oxygen demand (BOD) and fecal coliforms [8–11]. The WQI-NSF is calculated based on weights assigned to each parameter, according to a statistical survey conducted using the DELPHI technique, elaborated by 142 experts. The weights of each parameter are shown in **Table 2**.

In 1973, The Engineering Division of the Scottish Research Development Department began a research into the development of the Scottish Water Quality Index (WQI-SCO). Using the Delphi technique, and based on the WQI-NSF, 10 parameters were selected for the WQI-SCO with their respective weights: OD (0.18); BOD (0.15), free ammonia (0.12); pH (0.09); total oxidized nitrogen (0.08); phosphate (0.08); suspended solids (0.07); temperature (0.05); conductivity (0.06) and *Escherichia coli* (0.12). The sum of the weights at this index must be equal to 1. Two forms of calculation were then tested: the arithmetic form (Eq. 1) and the geometric form (Eq. 2), the second one being more efficient and less biased to high quality indices, and therefore more used. [9–12]

$$WQI = \sum wi * Si \tag{1}$$

$$WQI = \prod Si^{wi} \tag{2}$$

Si corresponds to the parameter and wi to its weight.

In Spain, one of the indices most accepted by the scientific community was proposed by Bascaron [13], represented in Eq. (3).

$$WQI = \frac{\left(\sum_{i=1}^{n} Ci * Pi\right)}{\sum_{i=1}^{n} Pi} \tag{3}$$

whereupon: n – number of parameters; Ci – value of the subscripts after normalization; Pi – assigned weight to each parameter.

The index proposed by Bascaron [13] presents normalization of the parameters seeking to balance the influence of each one on the final value of the index. In addition, it is a very flexible index in addition of other parameters [9, 13–15].

In order to represent the trophic status of a reservoir through WQI, Steinhart et al. [16] developed the Environmental Quality Index (EQI) for the Great Lakes of North America. This index was elaborated to collaborate with the multimillion-dollar cleaning projects of these lakes developed in the 1970s. The authors sought to evaluate nine physical-chemical, biological and toxicity variables. These variables were based on specific electrical conductivity, concentration of chlorine, pollutants with specific characteristics of odor and color, organic and inorganic toxic contaminants. The obtained data were converted into subscripts through mathematical functions which included national and international tolerance limits. In addition, these sub-indices were multiplied by weights assigned to each type of variable (0.1 for chemical (C), physical (P) and biological (B) parameters and 0.15 for toxic (T) substances). The quality assignment range varies from bad (0) to optimal (100). Each index, then, has its corresponding symbology indicating which parameter was more problematic, for example, a $70C_1P_1$ assessment means that a chemical parameter and a physicist are outside the limits stipulated by the legislation, even the quality attribution being considered good [9, 6].

In 1970, in Oregon, Canada, the O-WQI index was developed. This index was modified by Dunnette [17] for the evaluation of water in fishing regions in icy waters, and therefore very sensitive to high temperatures. This index became more used after Cude [18] re-evaluated it and added total phosphorus as an indication of the risk of eutrophication of the water body in Oregon [17, 18].

Cude [18], by proposing modifications, stated that the use of weights for the parameters which compose a WQI is only justifiable when used for evaluation of the water body for single use. Therefore, this author eliminated the weights of the variables and used a non-harmonic root equation. The parameters selected were DO, Fecal coliforms, pH, nitrate + ammonia, total solids, BOD, total phosphorus and temperature. The evaluation ranges defined for the index classified the water in excellent (90–100), good (85–89), acceptable (80–84), bad (60–79) and poor (10–59) [4, 14, 17–21].

The O-WQI index has been used in several parts of the world after the modifications proposed by Cude [18]. The calculation of the harmonic mean by the square root, Eq. (4), although it is more sensitive to the variations of the parameters, sometimes it can present ambiguity, as for example, the individual variables have good quality values and the average ones do not corroborate for this result of quality [4, 18–21].

$$O - WQI = \left[\frac{1}{n}\sum_{i=1}^{n} S_i^{-2}\right]^{-0.5} \tag{4}$$

whereupon Si – the value of the subscript corresponding to the variable under analysis; n – number of parameters.

Also in relation to Canada, another well-accepted and used index (WQI-CCME) is proposed by the Canadian Council of Ministers of the Environment [22]. The objective was to manage the quality of water by the water treatment and distribution agencies in the country. This is an index which allows a flexibility in the alteration of the variables, being suitable for the most diverse uses and morphoclimatic characteristics of the hydrographic basin under analysis. In addition, there is no need to evaluate subindices, nor weights for each variable. The aggregation model consists of a few steps. First, the variables must be standardized and three factors are determined (F1, F2, F3), where F1 refers to how many parameters that are not within the quality standards (Eq. (5)), F2 is the percentage of samples which has one or more non-standard parameters (Eq. (6)), F3 is calculated in three steps: the number of times the individual concentration of a parameter is outside the limit allowed by the law, called the excursion calculation (Eqs. (7) and (8)); after the calculation of the excursion, the sum of the excursion values is divided by the total number of tests (Eq. (9)); then F3 is calculated through Eq. (10). It should be noted that Eqs. (7) and (8) for the determination of the excursion are for concentrations of the parameters above and below the limits, respectively. Thus, to calculate the value of the WQI-CCME, we use Eq. (11).

$$F1 = \left(\frac{\text{number of variables out of standards}}{\text{total variables}} \right) \times 100 \tag{5}$$

$$F2 = \left(\frac{\text{number of tests out of limits}}{\text{total tests}} \right) \times 100 \tag{6}$$

$$excursion_i = \left(\frac{\text{number of times that concentration was below limits}}{\text{legislation limit}} \right) - 1 \tag{7}$$

or

$$excursion_i = \left(\frac{\text{legislation limit}}{\text{number of times that concentration was below limits}} \right) - 1 \tag{8}$$

$$nse = \left(\frac{\sum_{i=1}^{n} excursion}{\text{number of tests}} \right) \tag{9}$$

$$F3 = \left(\frac{nse}{0.01nse + 0.01} \right) \tag{10}$$

$$WQI - CCME = 100 - \left(\frac{\sqrt{F1^2 + F2^2 + F3^2}}{1.732} \right) \tag{11}$$

Some authors criticize the WQI-CCME for the calculation of F1, as it would not be adequate for a few variables (requires at least four) or with covariance among them [14, 20, 23].

The WQI calculation can be developed by several types of mathematical and statistical techniques. The indices, when created, tend to meet the needs of the region, the use of water and the parameters which can be measured. These criteria may bring some deviations in the calculation and, consequently, in the interpretation of the results. Smith [24] states that using, for example, a multiplicative WQI calculation method can lead the index to a low score if one of the variables has its individual score low. Thus, Smith [24] used an index based on a minimum operator in the aggregation of the subscript, according to Eq. (12).

$$WQI = Min(I_1, I_2, I_3, \ldots) \tag{12}$$

where upon: I, the sub-index of the *i*th parameter.

However, this type of WQI calculation ends up poorly reflecting the average quality since only one index will be attributed to the WQI, occurring the eclipsing phenomenon of the other variables. This index was adopted in New Zealand as a way of legislating and disseminating information about water quality [9, 14, 19, 21, 24].

The evaluation of WQI and its various forms of aggregation, the most commonly used being the harmonic mean, geometric, arithmetic and minimum function method mentioned above, should also take into account the financial question. The financial reality of the country, therefore, affects the choice of WQI. This factor must be observed, as this may reflect the amount of resources which will be allocated so that certain parameters can be better monitored. Methods such as the minimum function and the arithmetic method can eclipse and bring ambiguity to the final result and the financial investments destined to the improvement of certain parameters end up not bringing the desired solution. Experts affirm that there is a need for a better evaluation regarding the choice of the aggregation method and the weight criteria of the environmental variables.

Biological, limnological and ecological studies have been increasing in the last 40 years, and a revision of the proposed indices is necessary for either new variables to be included or their weights changed. In addition, there is a lack of indices which can better portray the impacts of climate change on the water body, as reported by Alves et al. [25] and that the improvement of the water quality indices would be fundamental. Alves et al. [25] point to the need for the development of indices that best portray tropical conditions such as those located in Brazil, since most of them were developed for temperate regions [4, 7, 18, 25, 26].

2.3. WQI for reservoirs

In Brazil, in 1975, the Environmental Company of the State of São Paulo (CETESB) started using the geometric quality index proposed by NSF (Eq. (2)), as shown in **Figure 1**. A number of studies developed in the country are based on this index [27–30]. In addition, a specific index for reservoirs was developed in Brazil and is widely accepted by the scientific community and the National Water Agency (ANA), which is the Water Quality Index for Reservoirs (IQAR), developed by the Environmental Institute of Paraná (IAP) [31–33].

The Environmental Institute of Paraná (IAP) seeking an evaluation of the water quality in reservoirs created this IQAR. The selected variables were dissolved oxygen deficit, total inorganic nitrogen, total phosphorus, chemical oxygen demand, transparency, chlorophyll a, weather, average depth and cyanobacteria. It should be noted that the phytoplankton community (diversity, algal bloom and amount of cyanobacteria) was included in the matrix through the concentrations of chlorophyll a and cyanobacteria, due to their ecological importance in

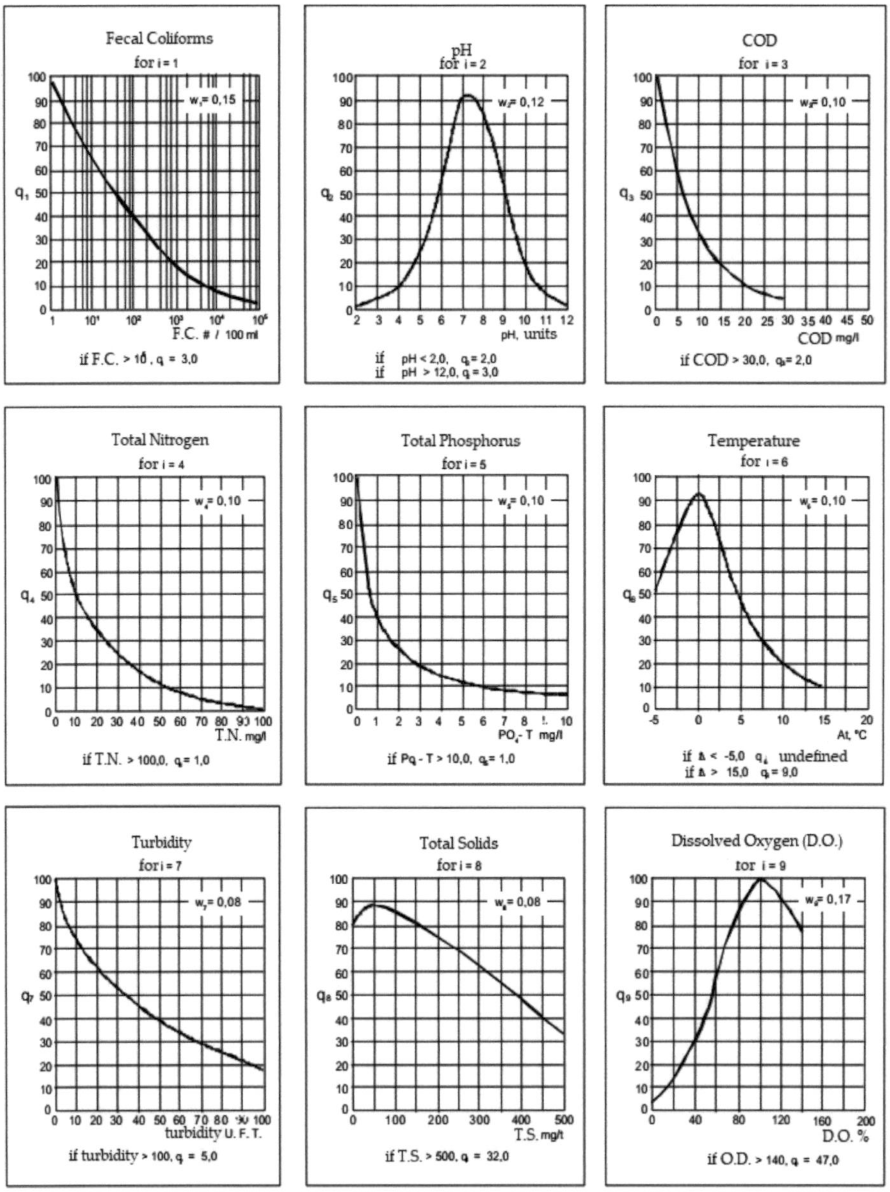

Figure 1. Average water quality variation curves of the WQI [33].

lentic ecosystems, however, this parameter received a different treatment. The developed matrix presents six classes of water quality, which were established from the calculation of the percentiles of 10, 25, 50, 75 and 90% of each of the most relevant variables and the selected parameters had weights assigned, as shown on **Tables 3** and **4** [34].

The water quality class to which each reservoir belongs is calculated through the Water Quality Index of Reservoirs (IQAR), according to Eq. (13).

Variables "*i*"	Weights w_i
Deficit of dissolved oxygen (%)*	17
Total phosphorus (mg L^{-1})**	12
Total inorganic nitrogen (mg L^{-1})**	08
Chlorophyll-a (μg L^{-1})***	15
Depth Secchi (m)	12
COD (mg L^{-1})**	12
Cyanobacteria (number of cells mL^{-1})**	08
Residence time (days)	10
Average depth (m)	06

*Average water column.
**Mean of depths I and II.
***Depth I.

Table 3. Selected variables and its respective weight [34].

Variables "i"	Class I	Class II	Class III	Class IV	Class V	Class VI
Deficit of oxygen (%)	≤5	6–20	21–35	36–50	51–70	>70
Total phosphorus (mg L^{-1})	≤0.010	0.011–0.025	0.026–0.040	0.041–0.085	0.086–0.210	>0.210
Total inorganic nitrogen (mg L^{-1})	≤0.15	0.16–0.25	0.26–0.60	0.61–2.00	2.00–5.00	>5.00
Chlorophyll-a (mg m^{-3})	≤1.5	1.5–3.0	3.1–5.0	5.1–10.0	11.0–32.0	>32
Depth Secchi (m)	≥3	3–2.3	2.2–1.2	1.1–0.6	0.5–0.3	<0.3
COD (mg L^{-1})	≤3	3–5	6–8	9–14	15–30	>30
Residence time (days)	≤10	11–40	41–120	121–365	366–550	>550
Mean depth (m)	≥35	34–15	14–7	6–3.1	3–1.1	<1
Phytoplankton (flowering)	No flowering	Rare	Occasional	Frequent	Frequent/permanent	Permanent

Table 4. Water quality matrix [34].

$$IQAR = \frac{\sum (q_i w_i)}{\sum w_i} \tag{13}$$

whereupon IQAR—Water Quality Index of Reservoirs; qi—water quality class in relation to the variable "*i*," which can range from 1 to 6; wi—calculated weights for the variables "*i*."

A partial IQAR is calculated from the data collected for each monitoring campaign. Then, the arithmetic mean of two or more partial indices are calculated to obtain the final IQAR and to define the quality classification of each reservoir. A reservoir is said to be impacted to much impacted if the IQAR varies from 0 to 1.5; the reservoir is poorly degraded when IQARs are between 1.51 and 2.5; moderately degraded when the IQAR values are between 2.51 and 3.5; the reservoir is considered to be degraded to polluted between 3.51 and 4.5; the reservoir is in the most polluted condition when IQAR has values between 4.51 and 5.5; and finally the reservoir is said to be extremely polluted when the IQA is equal to or above 5.51 [34].

Reservoirs are lentic environments that suffer greatly from the variation of temperature and oxygen by depth, leading to the appearance of several problems such as the process of eutrophication and anoxia in the deep parts, especially if this reservoir has organic matter in high concentrations and metals in its sediments. In this context, Lee et al. [35] sought to develop a WQI for reservoirs and lakes in South Korea, selecting parameters which would contribute to the solution of these problems and to describe the geology, climate and morphology of more than 70 lakes and reservoirs evaluated. The Water Quality Index for Lakes (LQI) was developed after the identification of the interrelationship of the parameters through the principal component analysis (PCA), and then four environmental parameters were selected: total organic carbon, chlorophyll a, total phosphorus and turbidity. Through the logistic regression, using a sigmoidal function for each parameter, the average LQI of the water body is calculated (**Table 5**). The LQI ranges from 0 to 100, in which 0 corresponds to a reservoir classified as poor quality and 100 as optimal quality [35].

Azevedo Lopes et al. [37], seeking to evaluate the quality of water for recreational purposes, proposed the Index of Conditions for Bath (ICB) to evaluate water bodies in Brazil, using the Delphi statistical technique to define the index composition variables: *E. coli*, density of cyanobacteria, visual clarity and pH. The authors used the minimal operator method, as originally proposed by Smith [24]. These authors affirm that the minimal operator method is effective for the definition of an index of quality and use of water bodies for recreational purposes, since it is enough that one of the variables present a low score in order to define the quality of the water body [24, 36, 37].

2.4. WQI through fuzzy logic and recent statistical techniques

Searching for improvement and accuracy in water quality indices, some computational techniques, such as fuzzy logic and neural networks, have been used. The fuzzy logic presented by

LQI equations for each parameter	Unit
$LQI(COT) = \left(1 - \frac{1}{1+78.9COT^{-3.11}}\right) \times 100$	mg/L
$LQI(Cl-a) = \left(1 - \frac{1}{1+24.8Cl-a^{-1.41}}\right) \times 100$	mg/m^3
$LQI(PT) = \left(1 - \frac{1}{1+583TP^{-1.93}}\right) \times 100$	mg/m^3
$LQI(Turb) = \left(1 - \frac{1}{1+17.5Turb.^{-2.02}}\right) \times 100$	NTU

Adapted by the author Lee et al. [35].
COT, total organic carbon; Cl-a, Chlorophyll-a; PT, total; Turb, turbidity.

Table 5. Lake quality index.

Zadeh [38] is widely used, mainly, in the environmental area in the issue of water quality, because it can avoid the ambiguity and the eclipsing effect of the variables. Several authors apply the fuzzy understanding to present the WQI with the smallest deviation and compare with the WQIs already developed. The selection of the variables does not obey a rigid criterion, and these can be chosen by the knowledge of the water bodies and the monitoring programs. Once the variables have been defined, they will be the input to the fuzzy system. The fuzzy inference system (FIS) consists then in the process of transforming these quantitative values into qualitative values (fuzzification) applying pertinence functions and established rules for the interaction between the parameters. Then, the inverse process (defuzzification) transforms these qualitative values into numerical ones (output). The accuracy of the use of fuzzy logic is related to the rules of interaction and the correct selection of parameters, which is a crucial step for the success of the index development [14, 20, 32, 38–42].

In Brazil, several pieces of research using the fuzzy logic for the development of water quality indices for reservoirs and rivers in the states of Rio de Janeiro, São Paulo, Sergipe and others have been developed over the years obtaining expressive results for WQI [32, 40–42].

In addition to the use of fuzzy logic, several indices have been proposed applying linear, nonlinear, multilinear regressions, principal component analysis (PCA) to identify weights, establish aggregation models and define variables which interfere with water body quality. These techniques for the development of water quality indices (WQI) are very recent in the area, showing at the same time a delayed insertion of this type of analysis in the improvement of WQIs, since many of the known indices were constructed using the statistical technique Delphi. The use of more sophisticated statistical techniques to reduce the size of the variables and to obtain more specific results in each analysis may be a way in the search for a single WQI which is possibly used worldwide. Although some authors defend this path, others believe that the need for an accepted model worldwide should be much more careful, especially regarding the use of the water body and the financial conditions of the country [9, 31, 43–45].

2.5. Future perspectives

The creation of new WQIs has been repetitive in terms of the aggregation models and the choice of parameters and, thus, basically without development of an index that is more adequate regarding the definition of the use and general applications. The use of less rigid, more objective formulas with flexibility in the choice of variables would contribute to the search for an accepted and efficient index worldwide. The construction of a more precise index is influenced by several factors, not only in terms of environmental parameters, but also in relation to the various aspects of environmental management. Thus, a method which would provide a kind of algorithm for the user to select the most coherent environmental variables would be very important. This selection should also be linked to the probable types of contamination of the water body in question, the financial aspects of the process for measuring these variables and the definition of the technical staff. Based on these considerations would be identified the use of water according to each legislation as well as the search for the appropriate water treatment would also be more efficient. In addition, the transmission of information would be faster between managers and society [4, 9, 19].

An important observation is the use of the WQI for agriculture which is still at an early stage, although works such as Stoner [46] present the division of the calculation for specific use of irrigation and public supply. Meireles et al. [47] used factor analysis (FA) and PCA to identify the variables which would most interfere with the amount of sodium in the soil, salinity and toxicity in the plants, in the Acaraú basin, Ceará. Out of 13 parameters evaluated by the authors, the ones which showed greater weight in the analysis were electrical conductivity, sodium, bicarbonate and Sodium Adsorption Index (SARo). The WQI was calculated by the sum of the quality product of each index multiplied by its respective weight and classified into five classes (**Table 6**) [9, 46–48].

Zahedi [49] sought to compare the index proposed by Meireles et al. [47] with the one developed in his work, through several statistical techniques, and to observe possible conflicts between the use of water from the wells for human supply and irrigation. This analysis aimed to identify water quality for both uses through WQIs. Much still has to be done to analyze conflicts over water use, but the indices can be a great support solution to these conflicts [46, 47, 49, 50].

Thus, the present work contextualized the importance of the introduction of indices related to traditional mathematical techniques alongside the most modern techniques such as fuzzy logic (FL), neural networks (NN) and machine learning (ML). Water management is an area of human knowledge which involves not only social and economic aspects, but also involves the visualization of using analytical technology to obtain environmental data and advanced computational technology, as presented on **Figure 2**.

WQI	Restrictions on water use	Recommendation	
		Soil	Plant
85–100	No restrictions (NR)	It can be used for most soils with low probability of sodification and salinization	No risk of toxicity to most plants
70–85	Low restriction (LR)	Recommended use for irrigated soils of fine texture or moderate permeability, being recommended the leaching of the salts. Salinization of thicker textured soils may occur and it is recommended to avoid use in soils with clay levels 2:1.	Avoid use in plants with salt sensitivity
55–70	Moderate restriction (MR)	It can be used in soils with high or moderate permeability, suggesting moderate salt leaching.	Plants with moderate salt tolerance still grow
40–55	High restriction (HR)	Can be used on soils with high permeability without layers of compaction. Programmed high-frequency irrigations can be adopted in water with electroconductivity above 2000 dS m^{-1} and SAR above 7.	It should be used to irrigate plants with moderate to high salt tolerance with special salinity control practices, except waters with low levels of Na, Cl and HCO_3
0–40	Severe restriction (SR)	Use for irrigation under normal conditions should be avoided. In special cases, it can be used occasionally. Soils with high salt water content have high permeability and excess water must be used to avoid accumulation of salts.	Only plants with high salt tolerance, except water with low levels of Na, Cl and HCO_3.

Table 6. WQI for irrigation [47].

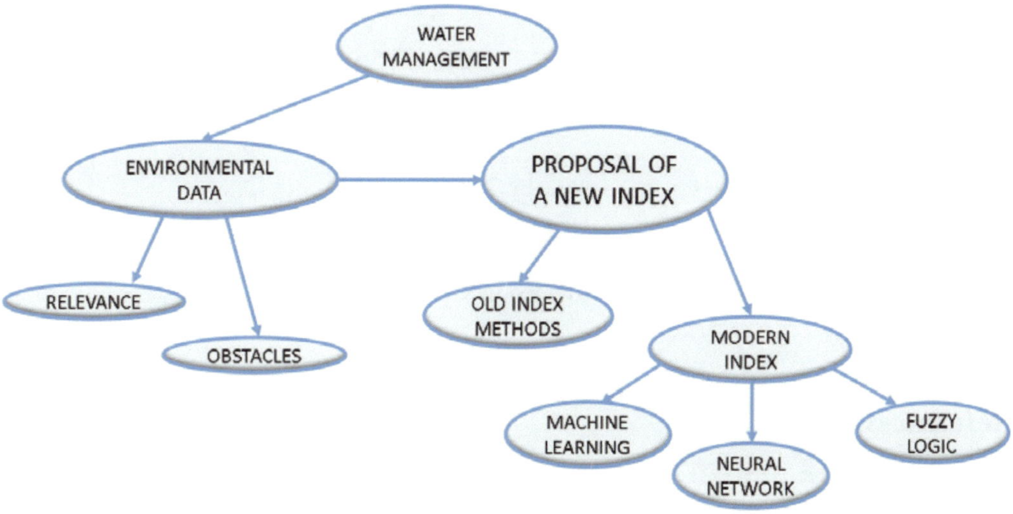

Figure 2. Monitoring and proposing water quality indices.

Environmental management models are composed of tools which evaluate natural phenomena and relate them to quantitative and qualitative values for decision making. Knowledge about the different parameters which characterize an environmentally degraded area is now one of the pillars of sustainable development and it is also a way of maintaining the sovereignty of a state in relation to the monitoring of water resources.

3. Conclusion

The evaluation of water quality by indices with several types of aggregation functions has been applied for many years. However, the use of these indices should always be linked to the area in which the water body is influenced, the climate of the region, the geology, the activities developed around it and its use by the population, so that it is possible to identify which physical, chemical and biological variables to be measured.

The type of the water body, lentic or lotic, will also influence this selection of variables for the construction of the index and another important factor which must be inserted in this one concerns the impacts of climate change. In this context, in order to meet these objectives, it is necessary to develop indices which are less influenced by eclipsing and ambiguity effects, requiring the use of better aggregation functions and statistics, as suggested by several authors.

In this sense, multivariate statistical techniques, regressions, machine learning and artificial intelligence, are now of relevance and importance for the creation of indices which encompass the most diverse variables for each specific use of water. In addition, this creation will facilitate decision making and communication between environmental managers and society. It is important to emphasize that such methods would allow a greater flexibility in the selection of variables, making water quality indexes more assertive, efficient, and easier to understand and manipulate.

Author details

Carlos Alexandre Borges Garcia[1]*, Igor Santos Silva[1], Maria Caroline Silva Mendonça[1] and Helenice Leite Garcia[2]

*Address all correspondence to: cgarcia@ufs.br

1 Universidade Federal de Sergipe, São Cristóvão, SE, Brazil

2 Chemical Engineering Department, Universidade Federal de Sergipe, São Cristóvão, SE, Brazil

References

[1] Horton RK. An index number system for rating water quality. Journal of Water Pollution Control Federation. 1965;**37**(3):300-305

[2] Derísio JC. Introdução ao controle da poluição ambiental. São Paulo: Editora: CETESB; 1992

[3] U.S. EPA. Environmental Impact and Benefits Assessment for Final Effluent Guidelines and Standards for the Construction and Development Category. Washington, DC: EPA Office of Water; 2009. EPA-821-R-09-012

[4] Walsh PJ, Wheeler WJ. Water quality indices and benefit-cost analysis. Journal of Benefit-Cost Analysis. 2013;**4**(1):81-105. DOI: 10.1515/jbca-2012-0005

[5] Sener S, Sener E, Davraz A. Evaluation of water quality using water quality index (WQI) method and GIS in Aksu River (SW-Turkey). Science of theTotal Environment. 2017;**585**: 131-144. DOI: 10.1016/j.scitotenv.2017.01.102

[6] OTT WR. Water Quality Indices: A Survey of Indices Used in the United States. Washington, DC: US Environmental Protection Agency; 1978. pp. 1-138. EPA-600/4–78-005

[7] Elshemy M, Meon G. Climate change impacts on water quality indices of the southern part of Aswan high dam reservoir, Lake Nubia. In: Fifteenth International Water Technology Conference, IWTC-15; 2011

[8] Brown RM, McClelland NI, Deininger RA, Tozer RG. A water quality index—Do we dare? Water & Sewage Works. 1970;**117**:339-343

[9] Lumb A, Sharma TC, Bibeault SJ. A Review of Genesis and Evolution of Water Quality Index (WQI) and some Future Directions. Water Quality Expo Health; 2011. pp. 11-24. DOI: 10.1007/s12403-011-0040-0

[10] Leitão VS, Cuba RMF, Santos LPS, Neto AS. Utilização do índice de qualidade de água (IQA) para monitoramento da qualidade de água em uma área de preservação ambiental. Revista Eletrônica em Gestão, Educação e Tecnologia Ambiental. 2015;**19**(3):794-803. DOI: 105902/2236117018330

[11] Misaghi F, Delgosha F, Razzaghmanesh M, Myers B. Introducing a water quality index for assessing water for irrigation purposes: A case study of the Ghezel Ozan River. Science of the Total Environment. 2017;**589**:107-116. DOI: 10.1016/j.scitotenv.2017.02.226

[12] Scottish Research Development Department. Development of a water quality index. In: Applied Research & Development Report Number ARD3. Edinburg, UK: Engineering Division; 1976. p. 61

[13] Bascaron M. Establishment of a methodology for the determination of water quality. Boletín Informativo Medio Ambiente. 1979;**9**:30-51

[14] Sutadian AD, Muttil N, Yilmaz AG, Perera BJC. Development of river water quality indices —A review. Environmental Monitoring Assessment. 2016. DOI: 10.1016/j.ecolind.2016.12.043

[15] Debels P, Figueroa R, Urrutia R, Barra R, Niell X. Evaluation of water quality in the Chillan river (Central Chile), using physical and chemical parameters and a modified water quality index. Environmental Monitoring and Assessment. 2005;**110**:301-322. DOI: 10.1007/s10661-005-8064-1

[16] Steinhart CE, Schierow LJ, Sonzogni WC. An environmental quality index for the Great Lakes. Water Resources Bulletin. 1982;**18**(6):1025-1031. DOI: 10.1111/j.1752-1688.1982.tb00110.x

[17] Dunnette DA. Oregon Water Quality Index Staff Manual. Portland, Oregon: Oregon Department of Environmental Quality; 1980

[18] Cude CG. Oregon water quality index: A tool for evaluating water quality management effectiveness. Journal of the American Water Resources Association. 2001;**37**(1):125-137. DOI: 10.1111/j.1752-1688.2001.tb05480.x

[19] Gitau MW, Chen J, Ma Z. Water quality indices as tools for decision making and management. Water Resources Management. 2016;**30**(8):2591-2610. DOI: 10.1007/s11269-016-1311-0

[20] Abbasi T, Abbasi SA. Water Quality Indices. 1st ed. Elsevier; 2012. 384 p. DOI: 10.1016/B978-0-444-54304-2.00001-4

[21] Swamee PK, Tyagi A. Describing water quality with aggregate index. Journal of Environmental Engineering ASCE. 2000;**126**(5):450-455. DOI: 10.1061/(ASCE)0733-9372(2000)126:5(451)

[22] CCME (Canadian Council of Ministers of the Environment). Canadian Environmental Quality Guidelines for the Protection of Aquatic Life, CCME Water Quality Index: Technical Report. Winnipeg, Manitoba: Canadian Council of Ministers of the Environment; 2001

[23] Tyagi S, Sharma B, Singh P, Dobhal R. Water quality assessment in terms of water quality index. American Journal of Water Resources. 2013;**1**(3):34-38. DOI: 10.12691/ajwr-1-3-3

[24] Smith DG. A better water quality indexing system for rivers and streams. Water Resources. 1990;**24**(10):1237-1244. DOI: 10.1016/0043-1354(90)90047-A

[25] Alves MTR, Teresa FB, Nabout JC. A global scientific literature of research on water quality indices: Trends, biases and future directions. Acta Limnologica Brasiliensia. 2014; **26**(3):245-253. DOI: 10.1590/S2179-975X2014000300004

[26] Poonam T, Tanushree B, Sukalyan C. Water quality indices—Important tools for water quality assessment: A review. International Journal of Advances in Chemistry. 2013;**1**(1): 15-28. DOI: 10.5121/ijac.2015.1102

[27] Medeiros SRM, Carvalho RG, Di Souza L, Barbosa, Da Silva AH. Índice de qualidade das águas e balneabilidade no Riacho da Bica, Portalegre, RN, Brasil. Ambiente & Água—An Interdisciplinary Journal of Applied Science. 2016;**11**. DOI: 10.4136/ambi-agua.1833

[28] Manoel LO, Carvalho SL. Avaliação Do Índice De Qualidade De Água (IQA) De Duas Nascentes No Município De Ilha Solteira-Sp. VII Encontro De Ciências Davida. UNESP, Campus de Ilha Solteira, Sp. Anais; 2013

[29] Gloria LP, Hornb C. Hidrográficas Através Da Ferramenta Do Índice De Qualidade Da Água – IQA. Revista Caderno Pedagógico, Lajeado. 2017;**14**(1). DOI: 10.22410/issn.1983-0882.v14i1a2017.1421

[30] Lima RS, Alves JPH. Avaliação da qualidade da água dos reservatórios localizados nas bacias hidrográficas dos rios Piauí-Real, utilizando o índice de qualidade da água (IQA). Scientia Plena. 2017;**13**:1-10. DOI: 10.14808/sci.plena.2017.109918

[31] Garcia HL, Mendonca MCS, Silva AF, Garcia CAB, Alves JPH, Costa SSL, et al. Assessment of water quality using principal component analysis: A case study of the açude da Macela. In: XVIth World Water Congress, 2017 Cancun-México. Anais do XVIth World Water Congress. 2017

[32] Silva IS, Mendonça MCS, Garcia CAB, Barreto VL, Garcia HL. Predição da qualidade da água do reservatório da macela através da lógica fuzzy. In: XIII Simpósio de Recursos Hidrícos do Nordeste. 2016

[33] CETESB, Companhia Ambiental do Estado de São Paulo. Apêndice D, Índice de Qualidade da Água [Internet]. 2017. Available from: https://cetesb.sp.gov.br/aguasinteriores/wpcontent/uploads/sites/12/2017/11/Ap%C3%AAndice-D-%C3%8Dndices-de-Qualidade-das-%C3%81guas.pdf [Accessed: 2018-09-17]

[34] IAP. Instituto Ambiental Do Paraná. Monitoramento da Qualidade da Água dos Reservatórios. Curitiba: IAP; 2016

[35] Lee Y, Kim JK, Jung S, Eum J, Kim C, Kim B. Development of a water quality index model for lakes and reservoirs. Paddy Water Environmental. 2014;**12**. DOI: 10.1007/s10333-014-0450-2

[36] Nagels JW, Davies-Colley RJ, Smith DG. A water quality index for contact recreation in New Zealand. Water Science and Technology. 2001;**43**(5):285-292

[37] Azevedo Lopes FW, Davies-Colley RJ, Von Sperling E, Magalhães AP Jr. A water quality index for recreation in Brazilian freshwaters. Journal of Water Health. 2016:243-254. DOI: 10.2166/wh.2015.117

[38] Zadeh LA. Fuzzy sets. IEEE Informatics and Control. 1965;(8):338-353

[39] Ocampo-Duque W, Schuhmacher M, Domingo JL. A neural-fuzzy approach to classify the ecological status in surface waters. Environmental Pollution. 2007;**148**(2):634-641. DOI: 10.1016/j.envpol.2006.11.027

[40] Lermontov A, Yokoyama L, Lermontov M, Machado MAS. River quality analysis using fuzzy water quality index: Ribeira do Iguape river watershed, Brazil. Ecological Indicators. 2009;**9**(6):1188-1197. DOI: 10.1016/j.ecolind.2009.02.006

[41] Garcia HL, Silva VL, Marques LP, Garcia CAB, Carvalho FO. Avaliação da qualidade da água utilizando a teoria fuzzy. Scientia Plena. 2012;**8**:1-10

[42] Pessoa MAR, De Azevedo JPS, Domingos P. Comparação das respostas de dois índices de qualidade de água usando dados simulados e reais. Revista Brasileira de Recursos Hídricos. 2015;**20**(4):905–913. DOI: 10.21168/rbrh.v20n4.p905-913

[43] Pathak H. Evaluation of ground water quality using multiple linear regression and mathematical equation. Analele Universităţii din Oradea, Seria Geografie. 2012;**2**(2):2-5

[44] Barakat A, Baghdadi ME, Rais J, Aghezzaf B, Slassi M. Assessment of spatial and seasonal water quality variation of Oum Er Rbia River (Morocco) using multivariate statistical techniques. International Soil and Water Conservation Research. 2016;**4**(4):284-292. DOI: 10.1016/j.iswcr.2016.11.002

[45] Tomas D, Curlin M, Maric AS. Assessing the surface water status in Pannonian ecoregion by the water quality index model. Ecological Indicators. 2017;**79**:182-190. DOI: 10.1016/j.ecolind.2017.04.033

[46] Stoner JD. Water quality indices for specific water uses. U.S. Geological Survey Circular 770. 1978

[47] Meireles ACM, Andrade EM, Chaves LCG, Frischkorn H, Crisostomo LA. A new proposal of the classification of irrigation water. Revista Ciência Agronômica. 2010;**41**(3):349-357. DOI: 10.1590/S1806-66902010000300005

[48] Almeida C, Quintar S, Gonzalez P, Mallea M. Assessment of irrigation water quality: A proposal of a quality profile. Environmental Monitoring Assessment. 2008;**142**:149-152. DOI: 10.1007/s10661-007-9916-7

[49] Cooper J, Rediske R, Northup M, Thogerson M, Van Denend A. Agricultural Water Quality Index. Allendale: Robert B Annis Water Resources Institute, Grand Valley State University; 1998. p. 75. WRIPublication # MR-98-1

[50] Zahedi S. Modification of expected conflicts between drinking water quality index and irrigation water quality index in water quality ranking of shared extraction wells using multi criteria decision making techniques. Ecological Indicators. 2017;**83**(July):368-379. DOI: 10.1016/j.ecolind.2017.08.017

Chapter 3

A Survey of Satellite Biological Sensor Application for Terrestrial and Aquatic Ecosystems

Mohamed E. El Raey and Hamdy A. Abo-Taleb

Additional information is available at the end of the chapter

http://dx.doi.org/10.5772/intechopen.85916

Abstract

The state of the ecosystems can be inferred in two ways, known as bioinference. One way (ground-based) is the use of some organisms to determine the environmental conditions within an ecosystem. The other is the use of multiband airborne or satellite imagery to identify the vegetation cover status, and also to track the biological diversity in marine ecosystems such as coral reef status, resources variation, and pollution. The standard example for the first state is the plankton as they represent a primary tool for ecologists to assess the health state of the marine environment. Their fast responses to the variability of the ecosystem, their nonexploitation as commercial organisms, and their favoring of subtle environmental conditions have suggested them to be bioindicators of climate variability. These organisms can be used to identify many environmental problems including water acidification, eutrophication, and pollution. Remote sensing technique is being widely used today to solve many environmental problems due to the broad view and accuracy of the results and its participation in determining the environmental conditions of different ecosystems. For example, remote sensing applications are used in vegetation and mangrove ecosystem management. Moreover, it is used to assess eutrophication problems by multiband spectrum remote sensing.

Keywords: bioindicators, remote sensing, plankton, mangrove, seagrass, aquatic and terrestrial

1. Introduction

New tools provide information that reduces the ever-increasing level of resources and cost now borne by regulatory agencies. Remote sensing tools, including satellite and airborne flights, as well as in situ devices, hold great promise for detecting selected taxa [1]. Alternative monitoring technologies and methods should be developed to enhance existing options used for regulatory purposes, further reducing the cost incurred by monitoring agencies. This should also include integration of new technologies into current monitoring programs, such as molecular probes, remote sensing data, and in situ instrumentation.

2. Remote sensing and land coverage

The estimation of vegetation cover utilizing remote sensing tools has become a primary method to gauge the impacts of regional and worldwide scale drought and agricultural status. It is likewise valuable in recognizing vegetation types in districts, including cultivated and wild assortments. Different investigations have exhibited that general biomass development can be estimated as well as biodiversity can be differentiated from the obtained data; additionally regular seasonal changes, like when blossoming happens and the effects of drought, and how dry season influences the associated species, could be estimated utilizing more incessant intra-seasonal imagery [2].

The application of different techniques of remote sensing to habitat monitoring, characterization of landscape, and geographical analyses of the cover change of the earth surface has made significant advances in the previous three decades [3]. However, at the least-developed areas, simple remote sensing tools are still relatively underused in ecological applications and especially in wetland systems and hydrobiology. Here, they have specific significance to inform environmental management system. There are several types of land cover changes that can be detected by using remote sensing applications that include agriculture cover thickening, regeneration and disturbance of vegetation, overgrazing, expansion of the urban, spatial changes in aquatic environment, and consequently the extent of the surface water, in addition to several changing processes of soil disturbance including accretion, abrasion, and erosion of the soil [4].

Turner et al. [5] checked out the application of remote sensing for monitoring biodiversity as well as conservation (reni application) over a wide spatial scale. Ecologists can use the help of the remote sensing applications in monitoring the vegetation state and detecting the changes in the environment and in areas where hard terrain, access difficulties, and extreme climate cause field studies very difficult. At spatial scales, remote sensing data are also appropriate for exploring the natural condition, type, biomass, geographical distribution, and productivity as well as quality of the vegetation ([4–8]). Classification of overall land use and change detection of cover [9–11] have worldwide applications particularly concerning assessments of the size and conversion rate of landscapes into urban and/or agricultural high-productivity systems. For aquatic ecosystems, data of airborne and satellite have a special use for monitoring changes of vegetation and water within these systems, as well as wetlands [12–14].

In Africa, remote sensing techniques have introduced valuable contributions in monitoring the environmental situation and diversity within multiple aquatic ecosystems [15–17]. Landsat information, for instance, has empowered the identification of the effects of both natural and human interaction on African wetlands, lakes, and freshwater eco-biological systems [18]. Results from the studies using information have uncovered that contamination coming about because of land-use changes, ecological modification, and different practices related with quick populace development increment and abstraction of water has caused or quickened many negative changes in the African lakes [19, 20].

The primary dangers to water quality in Africa are eutrophication, contamination, increasing water demand, and the expansion of invasive aquatic plant species like the water hyacinth [21]. NASA's earth observing system program studied Lake Chad; the results indicated that the lake declined to 1/20th of its original known size 35 years ago [22]. Abo-Taleb et al. [23] proved that the problems of eutrophication and pollution in Lake Idku have increased to the highest levels in addition to the shrinking of its total area and the depth of water in it. The Egyptian Lake Nasser development in North Africa and the new delta in the lake southern section have been monitored by remote sensing. Also the environmental changes of North African Lake Ichkeul in Tunisia were monitored [24, 25].

The use of remotely sensed data for monitoring North African wetlands has been effective [26] when combined with ground surveys and existing field data. An obvious example was the monitoring of environmental conditions and plankton distribution at the North African "El-Mex Bay" at the Egyptian Mediterranean Coast by Abo-Taleb et al. [27] by comparing in situ data with the satellite ones using GIS program. Turner et al. [5] focused on the significance of acquiring validator ground proof to empower to translate remote detecting items. For the administration of North African seaside ponds, remote sensing strategies give excellent methods for recognizing vegetation cover on a spatial scale and defining the open water extent, in addition to peripheral aggravation and quality of water [28, 29].

One of the most significant utilizations of remote sensing has been estimating agricultural yield through measurements of NDVI as this enables local and global organizations to assess what the outputs of yield will be. NDVI has been utilized to prognostic in advance of yield outcomes by standing on the developmental stage of agricultural vegetation and contracting it to the past. The usefulness of this permits sufficient time for drought problem-related choices and decisions to be made by relevant organizations and governments [30]. Techniques have been inserted to integrate imagery, symbolism, and statistical and simulation-based outcomes to foresee yields in various areas. Various sensitivities and seasonal changeability to natural conditions have made prediction challenging in some nations. Therefore, the best option might be using multiple techniques, where strategies could be developed more particularly for nations or areas that have more prominent opportunity for environmental variability [31].

Korets et al. [32] studied Northern Siberia boreal forests of Evenkia (~3600 km^2). An algorithm of forest cover mapping based on combined GIS-based analysis of ground truth data, digital elevation model (DEM), and multiband satellite imagery was developed. Using the classification principles and Kolesnikov approach, maps of forest growing conditions (FGC) and forest types were built. The resulted first map (**Figure 1**) was based on remote sensing (RS) composite classification, while the second one was developed by the digital elevation model (DEM) composite classification (**Figure 2**). The percentage of each forest component can result in **Table 1**.

2.1. Requirements

1. The images with very high spatial resolution (metric and sub-metric ones), obtained in mono or stereo mode, to identify agroforestry plot and tree crop structure, mixed cropping and intercropping, and agroecological infrastructures (riparian and hedgerow forests)
2. High image cloud-free obtaining recurrence of one to three on the biweekly sequence to distinguish practices whose recognition depends on the extraction of phenology-based features
3. Different consecutive croppings and so forth or practices of which the impact is of the brief term and needs many image acquisitions to be duly observed, for example, the date of the harvest and the mode, management of soil tillage, and residues of the crop

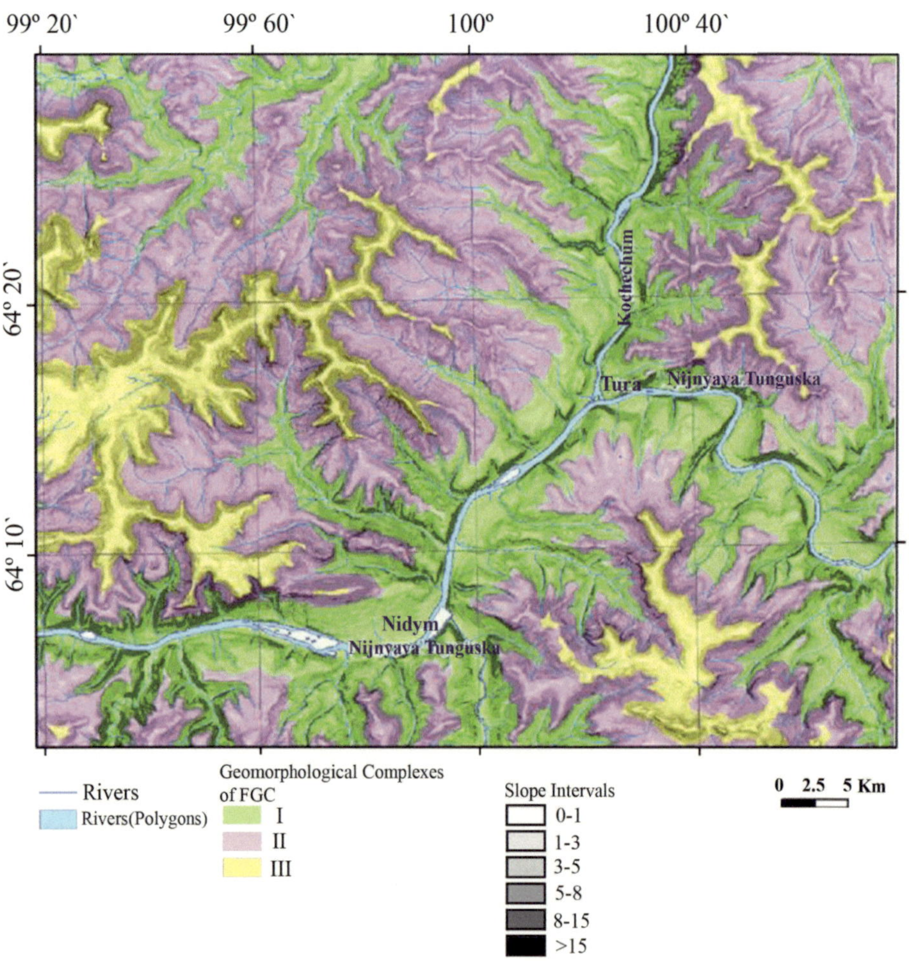

Figure 1. Two-layer composite map of potential forest growing conditions for two hierarchical classification levels: geomorphological (GMC) and types of forest growing conditions (FGC) [32].

http://dx.doi.org/10.5772/intechopen.85916

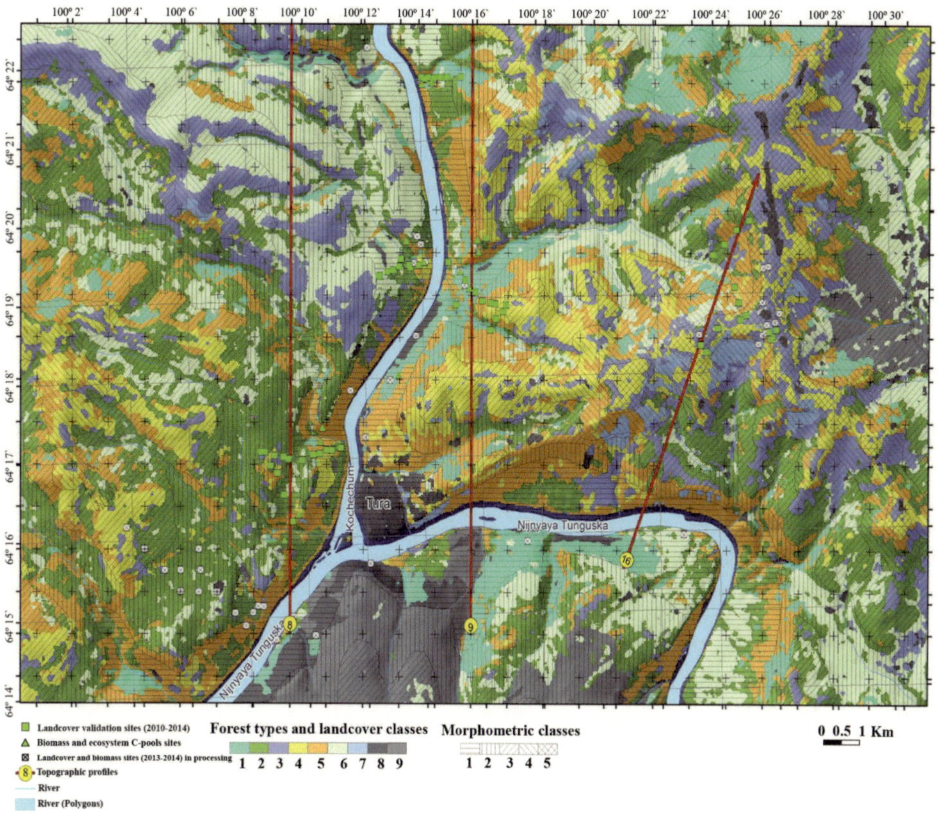

Figure 2. Remote sensing-based forest-type layer overplayed by digital elevation model (DEM-based) morphometric classes for the fragment of the study site [32].

Number	Types and other land cover classes	Area percent (%)
1	Larch stands, ledum, feather moss	10.6
2	Larch stands, dwarf shrub, lichen, feather moss	25.9
3	Birch stands, blueberry, sedge, feather moss	9.9
4	Larch stands, dwarf shrub, ledum, lichen	8.9
5	Larch open woodlands, feather moss, dwarf shrub, lichen	19.7
6	Larch stands, dwarf shrub, feather moss	16.7
7	Birch stands, sedge, feather moss	1.7
8	Open surfaces, rocks	2.9
9	Burned area of 2009	2.0
10	Another	1.6

Table 1. Percent of different forest components (class percent) of northern Siberia boreal forests of Evenkia [32].

4. Using hyperspectral images in crop variety identification
5. Airborne lidar information (light detection and ranging technologies) utilized for a 3D tree structure (tree harvests and hedgerows) and the characterization of soil culturing and tillage
6. The synthetic aperture radar (SAR) information utilized in cloudy regions for analysis of crop succession, multiple cropping mapping, and some product administration techniques

2.2. The encountered difficulties

1. Most of the research was done at the local scale, on small-size zones, and on a little number of fields, raising the issue of the cloning of the methodology and consequently the capacity to be developed at a local scale.
2. The small investigation areas are subjected to having a limited spectral variability because of environmental conditions and homogeneous practices, and at this scale, local information can be effortlessly utilized in the image interpretation process.
3. Remote sensing studies on the vegetation coverage need large training data sets that can be difficult to get to.
4. At the point when high-resolution images are utilized, the examinations are for the most part limited to small regions, because of the massive processing requests in time and hardware capacities.
5. The conversion of the satellite data to information about a cropping practice is generally completed through regulated picture classification, or statistical analysis, and often involves derived features such as spectral indices and, most commonly, vegetation indices.

By and large, most of the studies are discovering examinations, tried at a regional small scale with a main reliance on ground information, including one detecting sensor at the same time, and are building up on the knowledge of local conditions and data. The fundamental challenge in acquiring data from remote sensing at a local scale with high precision is the temporal as well as spectral changes of vegetation cover that is multifactorial. These changes or variability is connected to nature (soil compose, nature of the climate. Atmosphere, topography, and other factors), the cropping framework (of the area or field, soil type and plowing assortment, planting pattern, variety, and others), and the obtained image configuration (proportion of shadow to the sunlit pictures with high spatial resolution). Luckily, individual cases and exceptions to this general comment exist, and some great outcomes have been accomplished at the regional scale, like for harvest pattern or sequent cropping mode mapping. Be that as it may, these maps were created mainly for the vast areas and large-scale agricultural systems because of the utilization of coarse spatial resolution time series [33]. The Landsat image contains seven bands, with different characteristics and uses (**Table 2**).

http://dx.doi.org/10.5772/intechopen.85916

Band No.	Name	Wavelength (μm)	Characteristics and uses
1	Visible blue	0.45–0.52	Maximum water penetration
2	Visible green	0.52–0.60	Good for measuring plant vigor
3	Visible red	0.63–0.69	Vegetation discrimination
4	Near infrared	0.76–0.90	Biomass and shoreline mapping
5	Middle infrared	1.55–1.75	The moisture content of soil and vegetation
6	Thermal infrared	10.4–12.5	Soil moisture; thermal mapping
7	Middle infrared	2.08–2.35	Mineral mapping

Table 2. Thematic bands of NASA's Landsat satellite [34].

3. Mangrove status detection

Mangroves consist of intertidal flora and fauna found in the tropical and subtropical regions of the world. Mangrove forests mostly occur along the estuarine areas, where there is a uniform mixture of sea and river water. Mangroves play both protective and productive roles for the coastal community. Mangroves in the mudflats along the coastline reduce the impact of cyclones and tidal waves entering the mainland. The mangrove wetlands serve as spawning and nursery grounds for many economically important finfish and shellfish [35]. They prevent soil erosion and stabilize the coastline and also help in land building process by trapping sediments and suspended solids. Mangrove forests harbor many endangered fauna including saltwater crocodiles and many resident and migratory birds. Mangrove wetlands play an important role in enhancing the fishery production of the adjacent neritic waters by exporting organic and inorganic nutrients [36]. Mangrove plants are capable of surviving in the saline water environment through unique adaptations such as stilt roots, viviparous seeds, salt glands, salt-excluding mechanism, leathery leaves with thick cuticle, and pneumatophores [37].

Mangrove wetlands in India are more than 487,100 ha, of which 275,800 ha represents 56.7% of mangroves existing along the east coast, while 114,700 ha (23.5%) along the west coast and the remaining Indian mangrove 96,600 ha (19.8%) is found in the Andaman and Nicobar Islands. The extent and species diversity of mangrove wetlands in the east coast of India are more than the west coast due to a large number of east flowing rivers characterized by the presence of larger brackish water bodies and a complex network of tidal creeks and canals [38].

Mangrove ecosystems are undergoing widespread degradation due to a variety of human-induced stresses and factors such as changes in water quality, soil salinity, diversion of river water, sedimentation, and conversion of mangroves to other land-use practices like agriculture, aquaculture, and industrialization [39]. Mangroves are also degraded due geomorphological (topographic changes) and hydrological changes. Indiscriminate use of mangrove resources and clear felling of mangrove forests for catering the firewood requirement earlier were also responsible for the present degraded status. Collection of fish prawns, crabs, and mollusks is the major fishing activity apart from the collection of prawn juveniles for aquaculture [40].

This section portrays how remote sensing methods were utilized to study the impact of different woody coastal vegetations and mangroves as a defensive measure against the Indian Ocean Tsunami during 2004. Remote sensing makes it possible to do a comparison about pre- and post-tsunami pictures of huge areas [41]. An example on this case study is the coastal vegetation detection in multispectral remote sensing images for the 2004 Indian Ocean Tsunami, where Chouhan and Rao [34] selected the investigation site based on the already existed topographic maps and medium-resolution Landsat imagery. Determination criteria included significant banns reported, the existence of woody non-vegetated and vegetated shorelines, homogeneous bathymetry, and good coverage imagery before and after tsunami satellite. The before and after tsunami Ikonos and QuickBird images were analyzed and compared through the multispectral analysis and the visual interpretation of coastal vegetation before the disaster and after its harm. The outcomes were approved in the field. The investigation site covers around 20 km of coastline along the eastern shoreline of Tamil Nadu, India (see **Figures 4** and **5**). The Pichavaram mangrove is arranged in the northern piece of the investigation site. Whatever is left of the site contained shrimp and agriculture farms. The mangrove is associated with the Coleroon estuary in the south through various backwater channels.

Figure 3. Satellite images demonstrate the seriously harmed agricultural region southeast of TS Pettai: (a) before tsunami and (b) after tsunami [34].

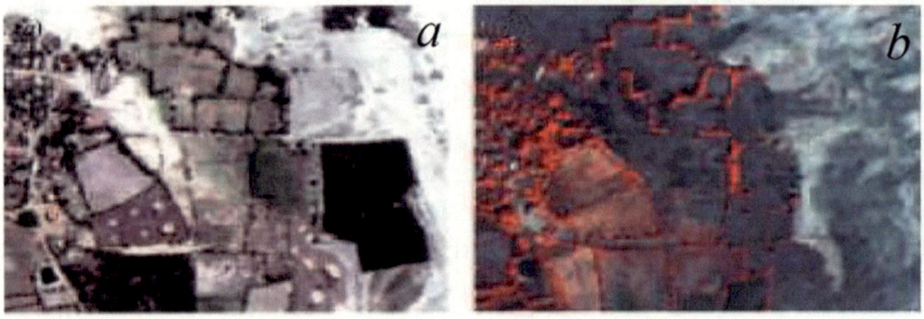

Figure 4. Images demonstrate the eastern piece of the whole village of Kodiyampalayam and the seriously harmed agricultural zones southeast of the village: (a) before tsunami and (b) after tsunami [34].

http://dx.doi.org/10.5772/intechopen.85916

Visual interpretation at **Figures 3, 4**, and **5** gives models of how the pre- and post-tsunami images were translated. **Figure 3** demonstrates the enormously harmed agriculture zone southeast of Pettai. The post-tsunami picture (**Figure 3b**) demonstrates that all highlights in the agrarian land have either vanished or are obscured contrasted with the pre-tsunami picture (**Figure 3a**). The multispectral interpretation showed that the image demonstrating the

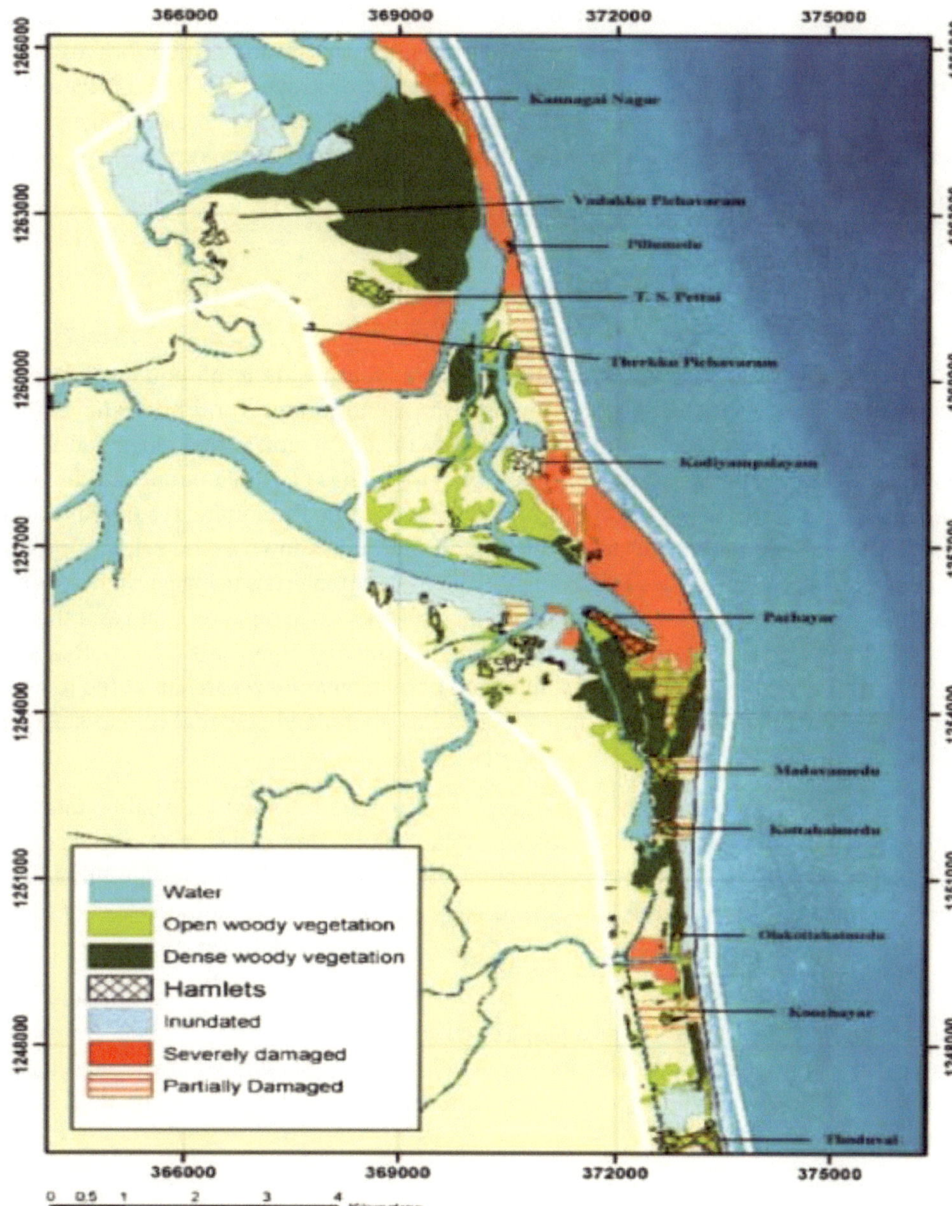

Figure 5. Map is delineating the damage degree and the territories protected by woody vegetation [34].

damage and the existence and nonattendance of the woody vegetation appears in **Figure 5**. The white contour shows the study zone, the light green alludes to open woody vegetation, the dark green alludes to thick woody vegetation, the red alludes to regions immensely harmed by the tsunami, the red-striped alludes to regions just in part harmed, and the spotted blue alludes to zones immersed by water, yet generally unharmed. The vast dim green region in the northern piece of the map is the *Pichavaram* mangrove.

Five villages are arranged close to this mangrove. Two of them are situated on the coast, while three villages are located to the west backward the mangrove. Ground truth information demonstrated that the two villages on the coast were totally destroyed, while the three villages behind the mangrove did not deteriorate by any means. Only north of these villages, territories at a similar distance from the ocean, yet without woody vegetative protection (defense wall), were immersed [34].

4. Remote sensing and aquatic system

For example, a large collection of aquatic organisms that reside in all aquatic ecosystems (oceans, seas, lakes, rivers) are known as plankton (aquatic animals and plants have limited powers for locomotion; they are under the mercy of the general water movement). These organisms are divided into two main groups, one of which is phytoplankton, and the second is zooplankton. They play a vital role in the food chain within the aquatic systems as they are located at the base of the food chain in the ocean water. It is known as the primary products that convert inorganic materials and elements into organic materials using sunlight in what is known as photosynthesis. It is transmitted to the higher levels in the food chain and therefore is the origin of life for the immature phases of all the aquatic organisms (like mother's milk for the terrestrial mammal) and is the stable food of many mature organisms at the top of the food chain such as some fish and whales. It also has a vital and essential role in stabilizing carbon in the ocean.

Photosynthesis represents a primary tool for the ecologists to assess the health state of the marine environment because of their role as bioindicators. Due to their fast response to the variability of the ecosystem, their nonexploitation as commercial organisms, and their favoring of environmental conditions, they have been suggested to be bioindicators of climate variability [42].

The presence and distribution—either vertically or horizontally—of these organisms are closely related with the surrounded aquatic environmental conditions as water salinity, temperature, viscosity, acidity, and density as the following equation in which depending on it the different water layer will be poor or rich with the organisms depending on their floating or sinking:

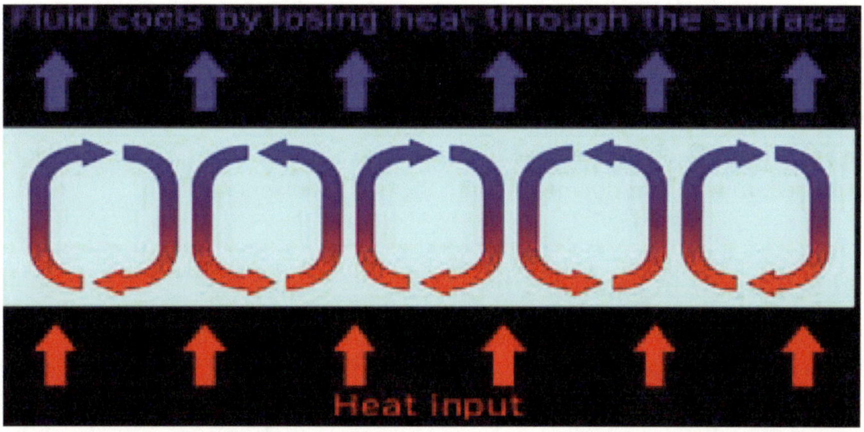

Figure 6. Heat convection cells. Source: Google search engine.

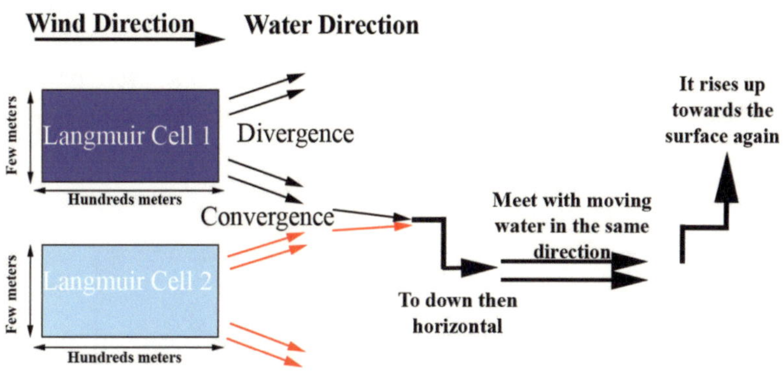

Figure 7. Langmuir convection cells.

On the other hand, these organisms can be used to indicate the direction of the water movement. This mechanism is related to the natural movements of the oceanic water layers (**heat convection** and **Langmuir convection** cells).

4.1. Heat convection

In the ocean the surface water heats up during the day and cools at night. These alternating heating and cooling change the density and create convection cells. Convection cells are the small units of water either sinking or rising according to their density. Plankton can utilize these gentle movements of the water particles to move up and down (**Figure 6**).

4.2. Langmuir convection cells

It results from the action of the wind blowing over the water and causes vertical movement of the water. It is produced when wind speed is above 3 m/sec. Each Langmuir cell is a few meters wide and hundreds of meters long. In each cell wind causes the water to move away from the center toward the outside (divergence). It meets the water from the adjusting Langmuir cell (convergence) where they go down along the line of convergence for a short distance and then move horizontally until they meet the water moving in the same direction from the adjusting cell where they rise to the surface again (**Figure 7**). The plankton organisms are carried upward and downward with the water movement.

5. Water acidification bioindication

In urbanized areas along the coastal zone, there is an obvious effect of water pH decreasing, and the ocean uptake of carbon from atmosphere increases the declining pH. The pH lowering is expected to continue in the next years. As the total inorganic carbon increases, the water depth is decreased. As a result, dissolving of calcium carbonate will be affected (because of the decreasing of water column in which the solubility occurs), causing a decline in surface water pH [43].

We can detect the water pH nature by monitoring the occurrence or absence of some aquatic groups. For example, pteropods (**Figure 8**) are characterized by a unique nature as they own shells composed mainly of aragonite which can significantly be dissolved in the acidic media. Hence, the absence of it is considered a strong signal and refers to the increase of the water acidity as recorded in hot spot areas of the Mediterranean Sea [42]. It is considered a highly sensitive organism to the environmental condition changes in pH.

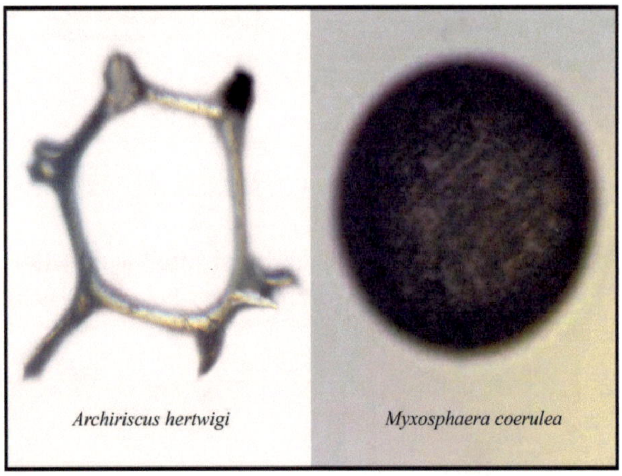

Figure 8. Some radiolarian species *Archiriscus hertwigi* and *Myxosphaera coerulea* which are very sensitive to water acidity rising *[42]*.

6. The water quality state biosensors

Some aquatic groups like ciliated Protozoa are considered as environmental state bioindicators. The disappearance of these organisms from any water body indicates the presence of

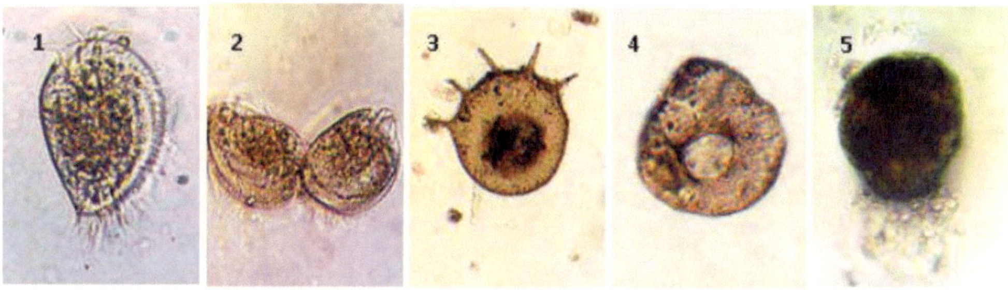

Figure 9. Some ciliate protozoan species. (1 and 2) *Euplotes* sp., (3) *Centropyxis aculeata,* (4) *C. ecornis,* (5) *Difflugia urceolata* [42].

toxic pollutants, such as cyanide, phenols, and heavy metals. On the other hand, the flourishing of these organisms indicates that the system is overloaded with oxygen deficiency and presence of putrefaction. The noticed increasing in the number of several different bacteria and the presence of Ciliata, *Cyanophyta,* and Zooflagellate are considered as an indication of aquatic system overloads with organic matter and an indication of oxygen deficiency and polysaprobic processes.

Some protozoan genera like *Euplotes, Centropyxis,* and *Difflugia* are considered as indicators on the pollution with sewage pathogens. These freshwater protozoan species are not encountered naturally in the marine water except when there is a freshwater discharging or sewage. Froneman, (2004) and Abo-Taleb et al. [42] reported that the presence of freshwater protozoan species in any marine coastal areas is considered as biomarkers on the presence of freshwater discharge into this region, and according to type of this species, we can determine the source of water discharged either rivers, drainage, lakes, or sewage (**Figure 9**).

Nematode organisms are sometimes encountered in the water column samples; the presence of these organisms is a sound cautionary signal on the contamination of the water body with sewage and final stage of putrefaction of organic matter. Additionally, it may be a signal on the pollution with the hydrogen sulfide.

7. Eutrophication biosensors

The words hypertrophic, eutrophic, mesotrophic, and oligotrophic have been used by scientists to describe the different nutritional statuses of the aquatic environment. The biologists use these words to describe the quantitative biomass which is potentially available.

Eutrophication is arising in the chemical nutrient salt content in the water, optimally compounds containing phosphorus or nitrogen, in an ecosystem. Consequently, it results in primary productivity increase in the ecosystem (excessive plant flourishing, growth, and decay) and further effects including oxygen deficiency and severe declining in water quality, fish stock, and other aquatic animal populations (**Figure 10**).

During the eutrophication, the concentration of the different nutrient salts in the water changes. Sometimes, one of the nutrients possibly linked to one of the aquatic organisms

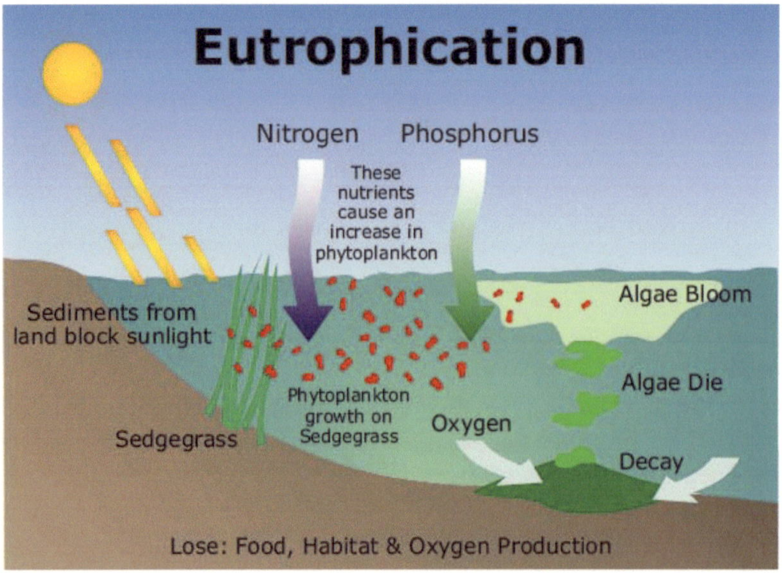

Figure 10. Eutrophication process and the main factors. Source: Google search engine.

excludes, so it will not be available for further algal growth. This excluded nutrient is called "the limiting factor." The nitrogen-to-phosphorus ratio in the water is an essential factor, and depending on it we determine which element of the two will be the limiting factor and consequently reduce the bloom [44]. Nitrogen is the limiting nutrient at many marine areas worldwide, especially during summer.

Some algal blooms, otherwise called "harmful algal blooms," are toxic to plants and animals. Toxic compounds they produce accumulate in shellfish and more generally in seafood, reaching dangerous levels for human and animal health.

The primary ocean productivity (plankton abundance) is primarily controlled by the fluctuations in several physical environmental conditions and nutrient concentration, which lead to high seasonality differences. Due to the eutrophication and pollution problems, many species thrive, while others become extinct. Without a doubt, for a clear understanding of the ecosystem, there is a necessity for a long-term monitoring data on the biological and physicochemical components.

The sensitivity of some groups, especially plankton species, to some chemical and physical conditions allows them to be used as biosensors or bioindicators of aquatic environment status. Being rather tolerant to different environmental conditions, a group like rotifers is a good indicator (biosensors) of the water quality due to its capability to tolerate severe environmental conditions especially the eutrophication and can be used for the environmental monitoring of the different water bodies.

With the progress of phytoplankton biomass increasing, the abundance of these primary producers causes herbivorous zooplankton organisms to abound. As a result of eutrophication and pollution, some species belonging to different groups like the copepod *Acartia clausi* is prevailed, while others became wade.

http://dx.doi.org/10.5772/intechopen.85916

8. Biosensors of fisheries

Amid the previous 40 years, the fishery efficiency of the world has been declining because of overfishing activities increasing, pollution, global warming, and climate change or natural surrounding change, contamination, and environmental change. Sustainable utilization of aquatic resources requires viable management, observation, and administration of the world's fish stocks. Remote sensing systems are being utilized to help in fishery sustainable management while additionally directing the fishing ships to the wealthy fishery ground and detecting the more effective fish shoal location. In the ocean, fishes tend to aggregate in some areas which have favorable conditions that change from one species to another; these conditions like primary productivity (watercolor), ocean surface temperature, and maritime fronts, which firmly impact common changes of fish stocks, Cannot be monitored and estimated by airborne and satellite remote sensors. The remotely detected information is provided in near-actuality time to help fishers save sailing time and fuel during their seeking for fish, modelers who offer fishery prognostications, and researchers who help evolve strategies for sustainable fishery administration [45].

Because of the human population increasing, overfishing, global environmental change, contamination, and natural surrounding corruption, around 40 years prior, sea productivity started declining, having achieved maximum sustainable yield. Almost 80% of the world fish stocks are wholly either currently exploited or overexploited [46]. Also, world interest for fish has been rising all over, both in developed nations because of rising standards of living and also developing nations, whose populace continues developing quickly [47]. Sustainable utilization of aquatic resources requires strict monitoring and management of whole ecosystems, not just abused fish stocks. Ordinary methodologies of sampling at the sea utilizing research vessels are restricted in both time and space scale of coverage, making it hard to think about the studying of the whole ecosystem. Since the beginning of satellite remote sensing, particularly remote sensing of sea surface temperature and color, it has become conceivable to sample the worldwide ocean on synoptic scales and with acceptable temporal resolutions [48–51].

There is also a wide range of practical fishery-related applications of remotely sensed data, including bycatch reduction, detection of harmful algal blooms, detection of fish shoal, aquaculture site selection, and identifying marine managed areas, as well as oceanographic and meteorological forecasting that improve scientific knowledge and safety of operations at sea [52, 53].

Discovering fish shoals and rich fishing sites is the fundamental reason for fuel consumption and vessel time cost in numerous commercial fisheries. To bring down the expense of fishing operations, there is a need to utilize biosensors, similar to a two-edged sword, which can be utilized not exclusively to help manage fisheries at sustainable levels yet, additionally, to guide fishing fleets to raise their catch. Early investigations demonstrated that satellite-determined fishery-help diagrams could lessen the search time of the US commercial fisheries up to 25–50%. Satellites can be utilized to find and anticipate prospective favorable zones of fish aggregation given the remotely sensed ecological indicators. These indicators may incorporate seafronts, separating waters of various colors or temperature;

upwelling zones, which are cooler and greener (more productive) than background waters; particular temperature ranges favored by certain fish; and so forth [45].

Laurs et al. [54] found that the catch rates of albacore—tuna that travels in large shoals and is of commercial importance as a food fish—were the maximum in blue warm oceanic waters, as satellite estimation correlated with the oceanfront (frontier between coastal and oceanic waters) (**Figures 11** and **12**). This image of satellite is one of a kind in that it displayed for the first time that remotely watched oceanographic features, like fronts, could be specifically concerning to fish catch. Shoreward interferences of oceanic water are synchronized with albacore aggregation zones. Laurs et al. [54] explain the gathering of albacore on the warmer side of the thermal fronts as a behavioral mechanism correlated with the feeding action, i.e., conglomeration of the tuna in clear water on the seaside of fronts in close shore zones mirrors a failure to proficiently get movable, large prey in turbid coastal water and a dependence on nourishment that moves over the oceanic coastal boundary.

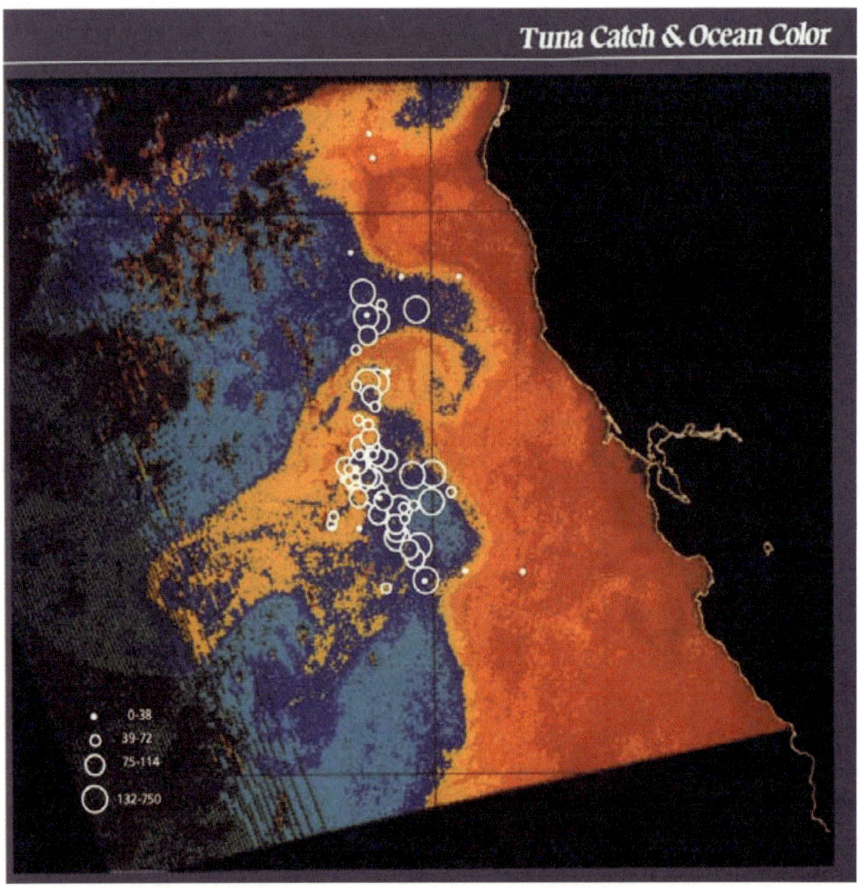

Figure 11. Nimbus-7 coastal zone color scanner (CZCS) satellite image showed locations of fish catch and their relation to the water color and showed a transition from coastal waters (orange color refers to the coastal water with high productivity, and blue color refers to the offshore areas). Source: NASA.

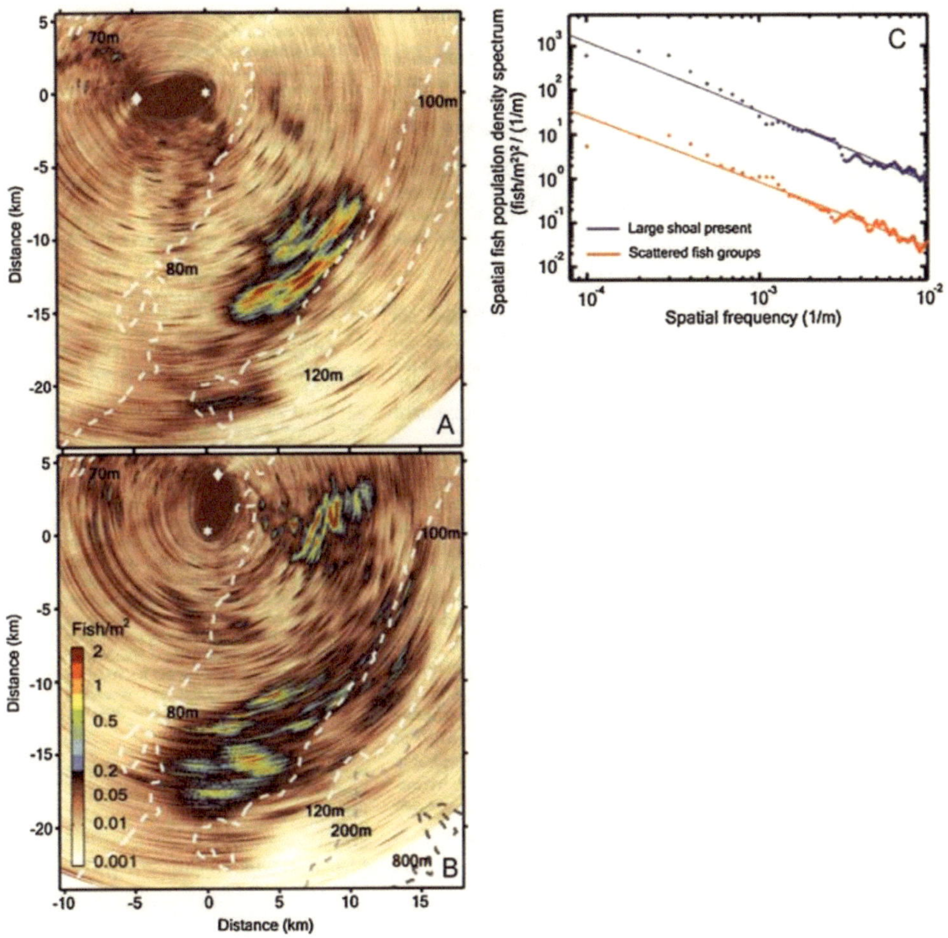

Figure 12. Two instant areal density images of fish shoals near the continental shelf edge obtained by ocean acoustic waveguide remote sensing on 14 May 2003 (A) and 15 May 2003 (B); and (C) is the spectrum analysis [55].

8.1. Airborne biosensors

Airborne biosensors are utilized to investigate the habitat of the fish and recognize fish shoals for the last 40 years. The main merit of using airborne remote sensing technique is that researchers can determine the remote sensing system characteristics. By picking the suitable focal length and flight altitude, they can steer the spatial resolution as well as the coverage. Moreover, the researcher can pick convenient atmosphere (like clear atmosphere without cloud), suitable tidal range (like the low tide), and sun angle ([56, 57]).

Drones are in particular cost-effective for coastal fish habitat detection nearshore. Trained atmospheric spotters have possessed the ability to detect menhaden, herring, and sardines' shoals, from low elevations. Skilled spotter pilots are used by fishing fleets to locate different fish shoals and direct the vessels by radio transmission [52, 58]. At night, fish shoals can be detected by the naked eye when plankton produces bioluminescence as a result of its stimulation by fish motions. To remedy this instance, a large number of airborne sensors have been added, including digital cameras, thermal infrared radiometers, low-light-level TV, and LIDAR and radar systems [52].

Airborne LIDAR (LIDAR is a system that emits laser light pulses that can penetrate up to three times the Secchi depth of a water column) has likewise been utilized to study coral reef, fish habitats and other sea life. An essential application of high-resolution imagers and airborne lidar is in coral reef fisheries, which is an area of significant source of income and food in developed and developing countries. Coral reef ecosystems are topographically complex environments, and this structural heterogeneity influences the behavior, abundance, and distribution of local ocean organisms. Satellite imageries, lidar, and high-resolution airborne images are being utilized to study and map these complex coral reef fish habitats and other ocean life [59–61]. Due to the strong relationship between the coral reef habitat and potential fish abundance and diversity, these maps are utilized by reef managers to facilitate ecosystem-based fishery management (EBFM) approaches, to guide sampling strategies, and to identify conservation areas.

Another useful airborne sensor is side-looking airborne radar (SLAR). Its operation depends upon emitting pulses and receiving signals that represent the backscattering intensities from the sea surface. The swimming fish close to the surface produces small-scale waves (2–20 cm length) which the radar can detect. The size and intensity of these wavelets rely upon school size, fish behavior at the surface, swimming activities, and fish size. The SLAR can pick up the little changes in the backscatter pattern caused by the fish shoal [52].

8.2. Advanced satellite remote sensing

Satellite images combined with other in situ data can be construed to find the suitable oceanic environmental conditions for fish aggregation [62]. Because certain species of game and commercial fish are indigenous to waters of a specific temperature and environmental conditions, fishers can spare ship tide and fuel by being capable of locating the higher potential sites more quickly [48, 63].

Other reason, that satellites are at the most modern and sophisticated in fisheries studies and resources management because the variability and magnitude of seas primary productivity that are very highly unknown on a vast worldwide scale, mainly due to the high temporal and spatial fluctuation of ocean phytoplankton abundance and diversity. As an example, in coastal regions, wind induced upwelling that conveys nutrients up to the water surface, causing patchy areas with high productivity, in addition to high chlorophyll and phytoplankton abundance, which can be monitored and detected by temperature and color sensors on satellites ([48, 64]: [65, 66]).

As appears in **Figure 13**, thermal infrared imagers and sea color sensors are utilized effectively to tracking coastal upwelling regions. In the upwelling areas in **Figure 13**, the water that ascends from the sea bottom conveying the nutrients upward to the sea surface appeared in the image of the thermal infrared as cool, while the sensor of sea color showed that the upwelling site as highly productive zone (the left image). Satellite sea color data was utilized to primary production measuring and fisheries status observing, as measuring the productivity of

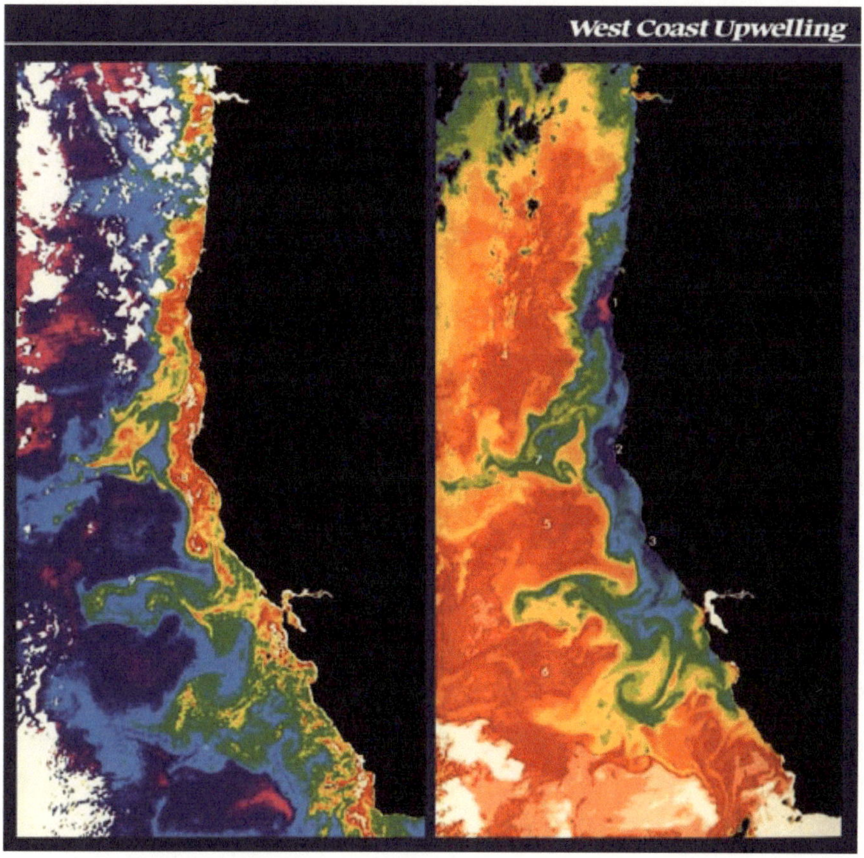

Figure 13. Satellite Ocean color and temperature maps (right) along the California coast showing the upwelling areas and chlorophyll distribution (left) along the California coast. Source: P. Zion and M. Abbott, jet Propulsion Laboratory, NASA.

Sensor	CZCS	SeaWiFS	MODIS Terra	MODIS Aqua	MERIS
Agency	NASA	NASA	NASA	NASA	ESA
Satellite	Nimbus-7	OrbView-2	Terra	Aqua	Envisat-1
Operating dates	1978–1986	1997–2010	Launch 1997	Launch 2002	Launch 2002
Spatial resolution (m)	825	110	250/500/1000	250/500/1000	300/1200
Number of bands	6	8	36	36	15
Spectral coverage (nm)	433–12,500	402–885	405–14,385	405–14,385	412–1050

Note that the MODIS instrument is carried aboard two platforms (Terra and Aqua).

Table 3. Some satellite remote sensing systems used to measure ocean color.

the four Eastern Boundary Currents (EBC) the Benguela, California, Humboldt, and Canary, as well as the currents over the first 24 months of Sea-Viewing Wide Field-of-View Sensor (SeaWiFS) operation. Inside every EBC, primary production has been assessed for dynamic zones of high chlorophyll concentration (more than 1 mg/m3) that showed levels of productivity probable to be used by higher trophic levels. Within every EBC the primary production diminished with latitude, while the range of the active zones is related to the volume of off-shore transfer. Differences in monitored fish catch were also correlated with the different trophic structure and spatial accessibility ([48, 52, 67, 68]).

The characteristics of some of the important satellite systems used to measure ocean color, such as Nimbus-7 Coastal Zone Color Scanner (CZCS), SeaWiFS, and Moderate Resolution Imaging Spectroradiometer (MODIS), are presented in **Table 3**.

9. Conclusions and recommendations

1. Newly developed satellite remote sensing techniques, combined with in situ measurements, constitute the most effective ways for efficient management and controlled exploitation of marine resources by combining in situ measuring data with satellite remote sensing ones.
2. Spectral and spatial resolutions of biosensors are the most important characteristics of the sensor.
3. Biosensors on board of satellites are capable of detecting and identifying conditions of mangrove and coral reef as well as water salinity, eutrophication, heat, and dynamics of fish shoals in the aquatic environment.

Author details

Mohamed E. El Raey[1]* and Hamdy A. Abo-Taleb[2]

*Address all correspondence to: melraey@gmail.com

1 Alexandria University, Alexandria, Egypt

2 Al-Azhar University, Cairo, Egypt

References

[1] Millie DF, Schofield OM, Kirkpatrick GJ, Johnsen J, Tester PA, Vinyard BT. Detection of harmful algal blooms using photopigments and absorption signatures: A case study of the Florida red tide dinoflagellate, Gymnodinium breve. Limnology and Oceanography. 1997;**42**:1240-1251

[2] Wang R, Gamon J, Montgomery R, Townsend P. Seasonal variation in the NDVI–species richness relationship in a prairie grassland experiment (Cedar Creek). Remote Sensing. 2016;**8**(2):128

[3] Schowengerdt RA. Remote Sensing: Models and Methods for Image Processing. New York: Academic Press; 2007

[4] Hall F, Botkin D, Strebel D, Woods K, Goetz S. Large-scale patterns of forest succession as determined by remote sensing. Ecology. 1991;**72**:628-640

[5] Turner W, Spector S, Gardner N, Fladeland M, Sterling E, Steininger M. Remote sensing for biodiversity and conservation. Trends in Ecology & Evolution. 2003;**18**:306-314

[6] Hobbs R. Remote sensing of spatial and temporal dynamics of vegetation. In: Hobbs R, Mooney H, editors. Remote Sensing of Biosphere Functioning. New York: Spring; 1990. pp. 203-219

[7] Franklin J. Land cover stratification using landsat thematic mapper data in Sahelian and Sudanian woodland and wooded grassland. Journal of Arid Environments. 1991; **20**:141-163

[8] Milne A, O'Neill A. Mapping and monitoring land cover in the Willandra Lakes world heritage region (New South Wales, Australia). International Journal of Remote Sensing. 1990;**11**:2035-2049

[9] Ambrose J, Shah P. The Importance of Remote Sensing and Mapping for Resource Management: A CaseStudy of Nepal. Integrated Surveys Section, Tomkcal Surveys Branch, Government of Nepal; 1990. pp. 2161-2164

[10] Petit CC, Lambin EF. Integration of multi-scale remote sensing data for land cover change detection. International Journal of Geographical Information Science. 2001;**15**:785-803

[11] Teng W. AVHRR monitoring of U.S. crops during the 1988 drought. Photogrammetric Engineering and Remote Sensing. 1990;**56**:1143-1146

[12] Hess LL, Melack JM, Novo EMLM, Barbosa CCF, Mary GM. Dual-season mapping of wetland inundation and vegetation for the Central Amazon basin. Remote Sensing of Environment. 2003;**87**:404-428

[13] Ozemi SL, Bauer ME. Satellite remote sensing of wetlands. Wetlands Ecology and Management. 2004;**10**:381-402

[14] Cózar A, Garcıa CM, Galvez JA, Loiselle SA, Bracchini L, Cognetta A. Remote sensing imagery analysis of the lacustrine system of Ibera wetland (Argentina). Ecological Modelling. 2005;**186**:29-41

[15] De Roeck ER, Verhoest NEC, Miya MH, Lievens H, Batelaan O, Thomas A, et al. Remote sensing and wetland ecology: A south African case study. Sensors. 2008;**8**:3542-3556

[16] Fuller RM, Groom GB, Mugisha S, Ipulet P, Pomeroy D, Katende A, et al. The integration of field survey and remote sensing for biodiversity assessment: A case study in the tropical forests and wetlands of Sango Bay, Uganda. Biological Conservation. 1998;**86**:379-391

[17] Munyati C. Wetland change detection on the Kafue flats, Zambia, by classification of a multitemporal remote sensing image dataset. International Journal of Remote Sensing. 2000;**21**:1787-1806

[18] UNEP. Africa's Lakes, Atlas of Our Changing Environment. Nairobi Kenya: United Nations Environment Programme; 2005

[19] Randazzo G, Stanley DJ, Di Geronimo SI, Amore C. Human-induced sedimentological changes in Manzala lagoon, Nile Delta, Egypt. Environmental Geology. 1998;**36**:235-258

[20] Scheren PAGM, Zanting HA, Lemmens AMC. Estimation of water pollution sources in Lake Victoria, East Africa: Application and elaboration of the rapid assessment methodology. Journal of Environmental Management. 2000;**58**:235-248

[21] Albright TP, Moorhouse TG, McNabb TJ. The rise and fall of water hyacinth in Lake Victoria and the Kagera River basin, 1989-2001. Journal of Aquatic Plant Management. 2004;**42**:73-84

[22] NASA. A Shadow of a Lake: Africa's Disappearing Lake Chad. Greenbelt, Maryland: Space Flight Center, National Aeronautics and Space Agency; 2001

[23] Abo-Taleb HA, Shaban WM, Hellal AM, Aboul Ezz SM, Sharaf MB. Assessing the ecological status of Edku Lake by using Rotifera as bio-indicators. Al-Azhar Bulletin of Science. 2017;**9**:235-249

[24] NASA. Lake Nasser and the New Valley. Greenbelt, Maryland: Goddard Space Flight Center, National Aeronautics and Space Agency; 2005a

[25] NASA. Ichkeul Lake, Tunisia. Greenbelt, Maryland: National Aeronautics and Space Agency: Goddard Space Flight Center; 2005b

[26] Flower RJ. Recent environmental change in north African wetland lakes: The CASSARINA project and the application of remote sensing for ecosystem monitoring. In: Fellous JL, editor. Satellite-Based Observation: A Tool for the Study of the Mediterranean Basin. Proceedings of an International Symposium at Tunis; 23-27 November 1998. Toulouse: Centre National d'Etudes Spatiale; 1998. pp. 219-224

[27] Abo-Taleb HA, Al Maghraby MA, El Raey M, Aboul Ezz SM, Abdel Aziz NE, Abou Zaid MM, et al. Mapping the different planktonic groups at one of the Egyptian bays along Mediterranean coast. Oceanography and Fisheries Open access Journal. 2018;**6**(5):1-12

[28] Ahmed MH, El-Leithy B. Utilization of satellite images for monitoring the environmental changes and development in Lake Mariout during the past four decades. In: Proceedings of the International Conference "Environment Is a Must"; 10-12 June 2008; Alexandria. 2008

[29] Ahmed MH, Donia NS. Spatial investigation of water quality of Lake Manzala using GIS techniques. Egyptian Journal of Remote Sensing and Space Science. 2007;**10**:63-86

[30] Kouadio L, Newlands NK, Davidson A, Zhang Y, Chipanshi A. Assessing the performance of MODIS NDVI and EVI for seasonal crop yield forecasting at the Ecodistrict scale. Remote Sensing. 2014;**6**(10):10193-10214

[31] Azzari G, Jain M, Lobell DB. Towards fine resolution global maps of crop yields: Testing multiple methods and satellites in three countries. Remote Sensing of Environment. 2017;**202**:129-141

[32] Korets MA, Ryzhkova VA, Danilova IV, Prokushkin AS. Vegetation cover mapping basedonremote sensing and digital elevation model data. The International Archives of the Photogrammetry, Remote Sensing and Spatial Information Sciences. 2016;**XLI-B8**:699-704

[33] Bégué A, Arvor D, Bellon B, Betbeder J, de Abelleyra D, Ferraz RPD, et al. Remote sensing and cropping practices: A review. Remote Sensing. 2018;**10**(99):1-32

[34] Chouhan R, Rao N. Vegetation detection in multispectral remote sensing images: Protective role-analysis of coastal vegetation in 2004 Indian Ocean tsunami. In: Geo-Information for Disaster Management Conference; Turkey. 2011. pp. 1-5

[35] FAO. Mangrove Forest Management Guidelines. FAO Forestry Paper 117; Rome; 1994

[36] Alongi DM, Boto KG, Robertson AI. In: Alongi DM, Robertson AI, editors. Tropical Mangrove Ecosystem. Washington, DC, USA: American Geophysical Union; 1992. pp. 251-292

[37] Ramasubramanian R, Ravishankar T, Sridhar D. Mangroves of Andhra Pradesh—Identification and Conservation Manual. Chennai: MS Swaminathan Research Foundation; 2003. p. 67

[38] Selvam V. Environmental classification of mangrove wetlands of India. Current Science. 2003;**84**:757-765

[39] Ravishankar T, Gnanapazham L, Ramasubramanian R, Sridhar D, Navamuniyamal M. Atlas of Mangrove Wetlands of India Part 2: Andhra Pradesh. Chennai: MS Swaminathan Research Foundation; 2004. p. 136

[40] Primavera J-H. A critical review of shrimp pond culture in Philippines. Reviews in Fisheries Science. 1993;**1**:151-201

[41] Olwig MF, Sørensen MK, Rasmussen MS, Danielsen F, Selvam V, Hansen LB, et al. Using remote sensing to assess the protective role of coastal woody vegetation against tsunami waves. International Journal of Remote Sensing. 2007;**28**(13):3153-3169

[42] Abo-Taleb HA, Aboul Ezz SM, Abdel Aziz NE, El Raey M, Abou Zaid MM. Detecting marine environmental pollution by biological beacons and GIS program. Journal of FisheriesSciences.com. 2016;**10**(4):069-083

[43] IPPC. Climate change 2007: The physical science basis. In: Solomon S, Qin D, Manning M, Chen Z, Marquis M, Averyt KB, et al., editors. Contribution of Working Group I to the Fourth Assessment Report of the Intergovernmental Panel on Climate Change. Cambridge: Cambridge University Press; 2007. pp. 1-996

[44] Abo-Taleb HA, El Raey M, Abou Zaid MM, Aboul Ezz SM, Abdel Aziz NE. Study of the physico-chemical conditions and evaluation of the changes in eutrophication-related problems in El-Mex Bay. African Journal of Environmental Science and Technology. 2015;**9**(4):354-364

[45] Klemas V. Fisheries applications of remote sensing: An overview. Fisheries Research. 2013; **148**:124-136

[46] FAO. The State of World Fisheries and Aquaculture 2008. Rome, Italy: FAO Documentation Group; 2009. p. 176

[47] APT. Depletion of Ocean Fisheries. Aquaculture Production Technology, Ltd. 2006. Available from: http://www.aquaculture.co.il/markets/deterioration.html [Accessed: October 4, 2011]

[48] Chassot E, Bonhommeau S, Reygondeau G, Nieto K, Polovina JJ, Huret M, et al. Satellite remote sensing for an ecosystem approach tofisheries management. ICES Journal of Marine Science. 2011;**68**:651-666

[49] Longhurst A. Mismanagement of Marine Fisheries. Cambridge, UK: Cambridge University Press; 2010. p. 334

[50] Stuart V, Platt T, Sathyendranath S. The future of fisheries science in management: A remote-sensing perspective. ICES Journal of Marine Science. 2011a;**68**:644-650

[51] Stuart V, Platt T, Sathyendranath S, Pravin P. Remote sensing and fisheries: An introduction. ICES Journal of Marine Science. 2011b;**68**:639-641

[52] Santos AMP. Fisheries oceanography using satellite and airborne remote sensing methods: A review. Fisheries Research. 2000;**49**:1-20

[53] Tyler JA, Rose KA. Individual variability and spatial heterogeneity in fish population models. Reviews in Fish Biology and Fisheries. 1994;**4**:91-123

[54] Laurs RM, Fiedler PC, Montgomery DR. Albacore tuna catch distributions relative to environmental features observed from satellite. Deep-Sea Research. 1984;**31**:1085-1099

[55] Makris NC, Ratilal P, Symonds D, Jagannathan S, Lee S, Nero RW. Fish population and behavior revealed by instantaneous shelf-scale imaging. Science. 2006;**311**:661-663

[56] Clark C, Ripley H, Green E, Edwards A, Mumby P. Mapping and measurement of tropical coastal environments with hyperspectral and high spatial resolution data. International Journal of Remote Sensing. 1997;**18**:237-242

[57] Myers JS, Miller RL. Optical airborne remote sensing. In: Miller RL, Del Castillo CE, McKee BA, editors. Remote Sensing of Coastal Aquatic Environments: Technologies, Techniques and Applications. Dordrecht, The Netherlands: Springer; 2005. pp. 51-68

[58] Lo NCH, Jacobsen LD, Squire JL. Indices of relative abundance from fishspotter databased on delta-lognormal models. Canadian Journal of Fisheries and Aquatic Sciences. 1992;**49**:2526-2551

[59] Andréfouët S, Gilbert A, Yan L, Remoissenet G, Payri C, Chancerelle Y. The remarkable population size of the endangered clam Tridacna maxima assessed in Fangatau atoll (eastern Tuamotu, French Polynesia) using in situ and remote sensing data. ICES Journal of Marine Science. 2005;**62**:1037-1048

[60] Hamel MA, Andréfouët S. Using very high resolution remote sensing for the management of coral reef fisheries: Review and perspectives. Marine Pollution Bulletin. 2010;**60**:1397-1405

[61] Purkis SJ, Graham NAJ, Riegl BM. Predictability of reef fish diversity and abundance using remote sensing data in Diego Garcia (Chagos archipelago). Coral Reefs. 2008;**27**:167-178

[62] McLain C, Hooker S, Feldman G, Bontempi P. Satellite data for ocean biology, biogeochemistry and climate research. EOS. Transactions of the American Geophysical Union. 2006;**87**:337-343

[63] Cracknell AP, Hayes L. Introduction to remote sensing; New York, Cram, DL: CRC press; 1977. Fishery surveillance and advisory systems in southern Africa. In: Tomczak GH, editor. Environmental Analyses in Marine Fisheries Research. Fisheries Environmental Services. FAO Fisheries Technical Paper No. 170. 2007. pp. 84-87

[64] Balch W, Evans R, Brown J, Feldman G, McLain C, Esaias W. The remote sensing of ocean primary productivity. Use of a new data compilation to test satellite algorithms. Journal of Geophysical Research. 1992;**97**:2279-2293

[65] Longhurst A, Sathyendranath S, Platt T, Caverhill C. An estimate of global primary production in the ocean from satellite radiometer data. Journal of Plankton Research. 1995;**17**:1245-1271

[66] Solanki HU, Dwivedi RM, Nayak R, Somvanshi VS, Gulati DK, Pattnyak SK. Fishery forecast using OCM chlorophyll concentration and AVHRRSST: Validation results off Gujarat coast, India. International Journal of Remote Sensing. 2003;**24**:3691-3699

[67] Carr M-L. Estimation of potential productivity in eastern boundary currents using remote sensing. Deep sea research part II: Topical studies in oceanography. Biological Sciences. 2001;**4**:959-980

[68] Aboul Ezz SM, Abdel Aziz NE, Abou Zaid MM, El Raey M, Abo-Taleb HA. Environmental assessment of El-Mex Bay, Southeastern Mediterranean by using Rotifera as a plankton bio-indicator. Egyptian Journal of Aquatic Research. 2014;**40**(1):43-57

Section 2

Air Quality Monitoring

Chapter 4

Atmospheric Aerosols Monitoring: Ground and Satellite-Based Instruments

Sunita Verma, Divya Prakash, Manish Soni and Kirpa Ram

Additional information is available at the end of the chapter

http://dx.doi.org/10.5772/intechopen.80489

Abstract

Aerosols are submicron particles suspended in the atmosphere which affect Earth's energy balance directly by scattering and absorbing the of solar radiation. In addition, they can indirectly affect radiation balance by changing the micro-physical and optical properties of the cloud. The difficulties in accessing the contribution of aerosols to radiative balance are caused partly due to incomplete knowledge of spatiotemporal variabilities in physicochemical and optical properties of aerosols on regional to global scale. Several state-of-the-art instrumentation techniques for ground-based measurements and satellite remote sensing technologies have been developed in past three decades to monitor physicochemical and optical properties of aerosols for a better understanding of radiative balance and feedback mechanisms. Satellite retrievals of moderate resolution imaging spectroradiometer (MODIS), ozone monitoring instrument (OMI), multi-angle imaging spectroradiometer (MISR) are used for this purpose. Ground-based measurements of aerosol properties provide a basis for validation of atmospheric correction procedures and can be used for validation of aerosol models used in atmospheric correction algorithms. This chapter describes in details about the widely used ground- and satellite-based remote sensing instruments for aerosol monitoring.

Keywords: aerosols, AOT, satellite, MODIS, particle, size, remote sensing, OMI, MISR

1. Introduction

Aerosols are a two-phase colloidal system, consisting of the particles (solid or liquid) and the gas in which they are suspended. An individual aerosol particle can either cause cooling or warming depending on its size, refractive index, composition and mixing state in the atmosphere

[1–3]. The interaction of aerosol particles with radiation further depends on their origins and subsequent atmospheric processing. The size of aerosols particles is an important parameter to study their chemical or optical properties. On the basis of particle size distribution the aerosols are broadly classified into three main categories: (1) Aitken particles or nucleation mode (0.001–0.2 μm diameter), (2) large particles, or accumulation mode (0.2–2 μm diameter) and (3) giant particles, or coarse particle mode (>2 μm diameter). Aerosol size distribution may be represented by number or volume density [4] (**Figure 1**). The particles of size less than 0.04 μm exhibit greater coagulation and hence their number is greatly reduced. Particles bigger than 10 μm are removed more effectively from the atmosphere either by sedimentation or by rain-wash.

Atmospheric aerosols have sizes, shape, and a chemical composition mainly due to their different emission sources and heterogeneous nature. Since the size, as well as composition of aerosol particles vary by orders of magnitude, no single instrument or technique is adequate for entire characterization of aerosols. The selection of a particular method for characterization of aerosols depends primarily on the particular type of application and scientific objectives. Aerosol optical thickness (AOT) is the most fundamental parameter for determining optical properties of aerosols which is the degree to which aerosols prevent the transmission of light in the atmosphere either through scattering or absorption. Measuring AOT at different spectral wavelengths (mostly visible range) helps in deriving information on the optical properties and to understand their impact on radiation balance while aerosol size distribution and mass concentration are crucial for understanding source strength, visibility and several environmental including fog-haze formation, cloud condensation nuclei (CCN), etc., and ecological impacts.

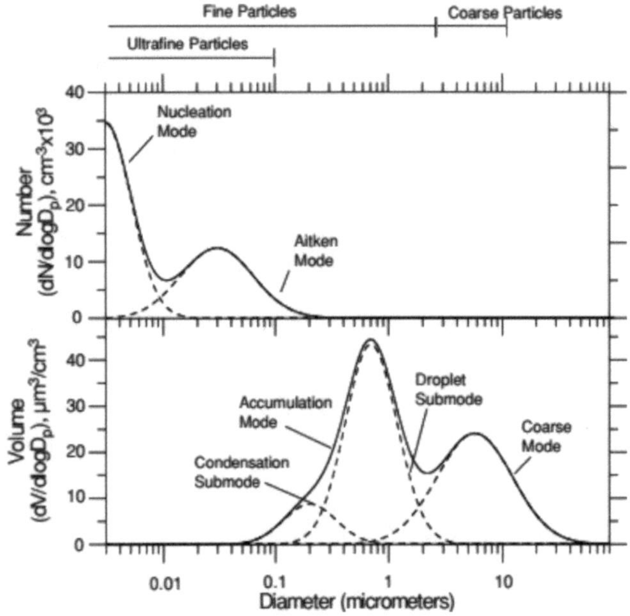

Figure 1. Typical volume and number distribution of atmospheric particles with different modes [4].

Three different kinds of data are available for the study of aerosols: (1) point data from ground locations equipped with sunphotometers (highest accuracy, but periods of missing data are frequent) [5]; (2) gridded data from space observations (low accuracy, with frequent missing data) [6]; and (3) gridded data from monthly climatologies (low to intermediate accuracy, no missing data) [6, 7].

Ground and satellite-based remote sensing are two important ways for monitoring aerosol optical properties [8]. The ground-based and satellite aerosol optical measurements contribute to our understanding of the Earth's systems by not only characterizing the ambient aerosol but validate satellite retrievals and numerical modeling algorithms. This also provides information on environmental pollution and investigates aerosol and cloud effects on radiative fluxes [8, 9].

More importantly, they provide a wider spatial as well as temporal (e.g., long-term variability) coverage of aerosol properties on a regional to global scale. Nonetheless, the resolution of measurements is rather poor for local observations and thus, these methods do not provide adequate information over a given site. In contrast, real-time measurements of chemical and optical properties are advantageous to study several physical and chemical processes on a finer spatial scale but fail to provide information about these processes on a regional scale. Therefore, a blended approach of simultaneous real-time ground-based measurements in tandem with ground and satellite-based remote sensing techniques is a need for better understanding of aerosol properties and their impact on the environment, ecology weather and climate.

The present chapter discusses several established as well as recent instruments in the field of atmospheric aerosol measurements with a particular focus on the measurements of aerosol optical depth and size. The brief description of instruments for aerosols along with their measurement techniques and working principles are discussed in details.

2. Ground-based aerosols instruments

Different types of ground-based instruments are used for monitoring aerosols such as spectroradiometers, multichannel radiometers, and broadband radiometers. Spectroradiometers present today provides excellent spectral performance across a full range of solar irradiance spectrum. Though, they are very expensive and thus, most ground-based stations generally prefer to deploy multichannel and broadband radiometers. Data from the ground are mostly used as ground truth to validate and optimally combine data from the other sources. Ground-based multi-wavelength sunphotometric measurements are the most accurate source of AOT data [9].

In multichannel and broadband radiometers, the wavelength integrated quantities are reproduced as accurately as in the spectroradiometers. This section further discusses in details about the ground-based aerosols measurement instruments. The brief description of sunphotometer for AOT and aerosol spectrometer for size distribution measurements are described in Sections 2.1 and 2.2.

2.1. Sunphotometer

Sunphotometer is capable of measuring AOT (AOT) and direct solar irradiance in different wavelength bands generally in 6 narrow spectral bands ranging from 360 to 1000 nm. It uses the Langley method for deriving optical thickness of the atmosphere by measuring direct sun irradiance. A correction for Rayleigh scattering as well as selected atmospheric gases such as ozone, water-vapor, etc., is needed. After subtraction of the Rayleigh optical thickness and the ozone optical thickness, the aerosol optical thickness is obtained. In the following section, we describe MICROTOP II sunphotometer as an example for sunphotometers.

2.1.1. MICROTOP II

Sunphotometer MICROTOP II (Model 540) is a five-channel small portable, handheld sunphotometer manufactured by solar light [10]. This is particularly useful for taking ground-based measurements at different locations (**Figure 2**). This instrument can give total column aerosol optical thickness spectra together with total column ozone content and precipitable water vapor within the atmosphere. The instrument is equipped with five accurately aligned optical collimators, with a full field of view of 2.5°, while the full width at half maximum (FWHM) bandwidth at each of the AOT channels is 2.4 ± 0.4 nm. It is a Voltz-type radiometer and measures the intensity of direct solar irradiance at five narrow-band spectral channels centered at 440, 500, 675, 870 and 936 nm. Each channel is fitted with a narrow-band interference filter and a photodiode suitable for particular wavelength range. The collimators are encapsulated in a cast aluminum optical block for stability. A sun target and pointing assembly is permanently attached to the optical block and laser-aligned to ensure accurate alignment with the optical channels. When the image of the sun is centered in the bulls-eye of the sun target, all optical channels are oriented directly at the solar disk. Radiation captured by the collimator and bandpass filters radiate onto the photodiodes, producing an electrical current that is proportional to the radiant power intercepted by the photodiodes. These signals are first amplified and then converted to a digital signal by a high-resolution A/D converter. The signals from the photodiodes are processed in series. However, with 20 conversions per second, the results can be treated as if the photodiodes were read simultaneously [11].

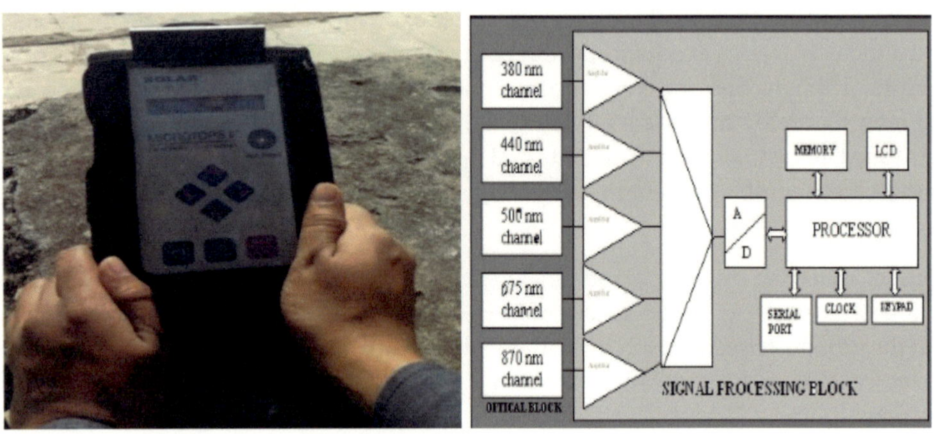

Figure 2. Operational block diagram of MICROTOP II [12].

Optionally the MICROTOP is capable of deriving the water vapor column. AOT and water vapor column are determined to assume the validity of the Bouguer-Lambert-Beer law. At a wavelength λ, the total optical thickness recorded by a sunphotometer can be expressed as the sum.

$$\tau_{\lambda} = \tau_{A\lambda} + \tau_{R\lambda} \tag{1}$$

where $\tau_{A\lambda}$ is AOT and $\tau_{R\lambda}$ is Rayleigh optical thickness.

The MICROTOPS II calculates the AOT (τλ) value of each wavelength based on the channel's signal, its extraterrestrial constant, atmospheric pressure (for Rayleigh scattering), time and location. Solar distance correction is automatically applied. The AOT formula is

$$\tau_{\lambda} = \frac{\ln(V_{0\lambda}) - \ln(V_{\lambda} \times SDCORR)}{m} - \tau_{R\lambda} \times \frac{P}{P_0} \tag{2}$$

where the index λ references the channel's wavelength, $\ln(V_{0\lambda})$ is the AOT calibration constant, V_{λ} is the signal intensity in mV, SDCORR is the mean Earth-Sun distance correction, m is the optical air mass, $\tau_{R\lambda}$ is the Rayleigh optical thickness and P and P0 are station pressure and standard sea level pressure (1013.25 mB) respectively [9].

Young [13] has given one approximate formula to calculate air mass (m) from true solar zenith angle (θ) that accounts for the effects of refractive index and curvature of the Earth.

$$m = \frac{1.002432 \times \cos^2(\theta) + 0.148386 \times \cos^2(\theta) + 0.0096467}{\cos^3(\theta) + 0.149864 \times \cos^2(\theta) + 0.0102963 \times \cos(\theta) + 0.000303978} \tag{3}$$

where θ represents the solar zenith angle at the time of measurement. The optical thickness due to Rayleigh scattering is subtracted from the total optical thickness to obtain AOT. Optical thickness from other processes such as O_3 and NO_2 absorption are ignored in MICROTOPS II. The combined error in the estimated AOT due to errors like entering of the diffuse radiations, computational error calculating the air mass, calibration coefficient estimation error and uncertainty in deriving optical depth due to Rayleigh scattering and absorption is in the range of 0.009–0.011 at different wavelengths (which is 2–10% of the total AOT). Typical errors in AOT measurements from MICROTOP-II are ~0.03 [14].

In 1961, Scientist named Angstrom proposed an empirical formula to estimate the spectral dependence of atmospheric extinction (scattering and absorption) caused by aerosols:

$$\tau_{\lambda} = \beta^{*} \lambda^{-\alpha} \tag{4}$$

where τ_{λ} is the spectral AOT, λ is the wavelength in μm; α is the Ångström exponent and β is the turbidity coefficient (equivalent to AOT at 1 μm) provides a measure of columnar aerosol loading. The values of α and β were calculated by evolving a linear least square fit between τλ and λ (in μm) in log-log scale over the entire wavelength range. The Ångström exponent (α) is a measure of the relative dominance of small particles while the coefficient β depends on the aerosol loading associating more with coarse aerosols.

2.1.2. CIMEL sunphotometer

The CIMEL (CE318) sunphotometer is also a multi-channel, automatic sun-and-sky scanning radiometer that measures the direct solar irradiance and sky radiance at the Earth's surface (**Figure 3**). The CIMEL sunphotometer is developed by French company Cimel Electronique [15].

CIMEL works on Rayleigh scattering principle and measures the total aerosol load in the atmosphere. AERONET (AErosol RObotic NETwork) is an optical ground-based aerosol monitoring network and data archive consisting of sunphotometers. This network provides globally distributed near real-time observations of aerosol spectral optical depths, aerosol size distributions, and precipitable water in diverse aerosol regimes. Measurements through CIMEL are taken at pre-determined discrete wavelengths in the visible and near-IR (i.e., at 340, 380, 440, 500, 675, 870, 1020 nm) parts of the spectrum to determine atmospheric transmission and scattering properties. The CIMEL sunphotometer takes measurements of the direct sun and diffuse sky radiance with 1.2° full field of view within the spectral range 340–1020 nm [16].

The direct sun measurements are done for all eight spectral channels with triplet observations per wavelength and sky radiance measurements at following four spectral channels (440, 675, 870, and 1020 nm). The instrument performs direct spectral solar radiation measurements within a 1.2 full field of view every 15 min. This data set is in the form of Level 2.0 quality assured product after cloud screening and necessary post calibration. The AOT is retrieved at all channels [16] other than the 940-nm channel, which is used to retrieve atmospheric water vapor content. The CIMEL sky radiance measurements together with the direct sun measurements of optical depths are used to retrieve optical equivalent aerosol size distributions and refractive indices. This instrument is weather-proof and requires little maintenance during periods of adverse weather conditions. It takes measurements only during daylight hours (sun above the horizon). It provides the quantification and physical-optical characterizations of the aerosols. CIMEL fully meets the operational requirements of continuous monitoring in terms of reliability, long lifetime and very low maintenance cost The large range of parameters that are derived and calculated from the measurements and from the atmospheric physics equations make the CE318 photometer a worldwide benchmark device for many applications like characterization and quantification of aerosols and aerosol types, bias correction for satellite aerosol retrievals,

Figure 3. The CIMEL Sunphotmeter instrument.

detection of volcanic ash plumes in real time, deriving parameters (i.e., aerosol optical depth, fine mode AOD, coarse mode AOD, optical properties), volume size distribution, air quality monitoring, etc. The types of error that instrumentation of sunphotometer faces are described in [17–19] These studies reveals that, the overall error in the AOT can be due to (a) error associated with the uncertainty in the optical thickness due to Rayleigh scattering and absorption by O_3 and water vapor (b) deviation of the calibration coefficient with time (c) diffuse radiation entering the optical channel and (d) computation error in relative air mass (a geometrical term to account for the relative increase in optical path length as solar zenith angle increases).

2.2. Aerosol spectrometer

The aerosol spectrometer is a device that is used for continuous measurement of airborne particles as well as for measuring the particle's number and volume concentration.

2.2.1. GRIMM aerosols spectrometer

The GRIMM (model 1.108) is a compact portable instrument developed by GRIMM Aerosol Technik Ainring contains an integrated gravimetric filter on which all particles are collected after the optical measurement [20]. GRIMM aerosol spectrometer measures the number of particles per unit volume of air using light-scattering technology. This instrument operates in four modes: count distribution, mass distribution, occupational health and environmental mode. The instrument measures particle concentrations in an optical size of 0.3–20 μm in 15 channels with differently sizes with a concentration range of 1–2,000,000 particles/L (for count distribution mode) or a mass concentration range of 1–1,00,000 μg/m^3 (for mass distribution, environmental and occupational health modes) [21]. The sensitivity of the spectrometer is 1 particle/L or 1 μg/m^3, whereas reproducibility of the instrument is ±2%. Ambient air is drawn into the unit via an internal volume-controlled pump at a rate of 1.2 L/min. The instrument initiates a system self-test (≈30 s) and zero calibration check at the start of each measurement. Afterward the actual measurement starts and the LCD-display shows data continuously every 6 s. A stainless-steel tube is utilized as the spectrometer inlet (**Figure 4**).

2.2.1.1. Measuring principle

The sample air is led directly into the measuring cell via the aerosol inlet. The particles in the sample air are being detected by light scattering inside the measuring cell. The scattering

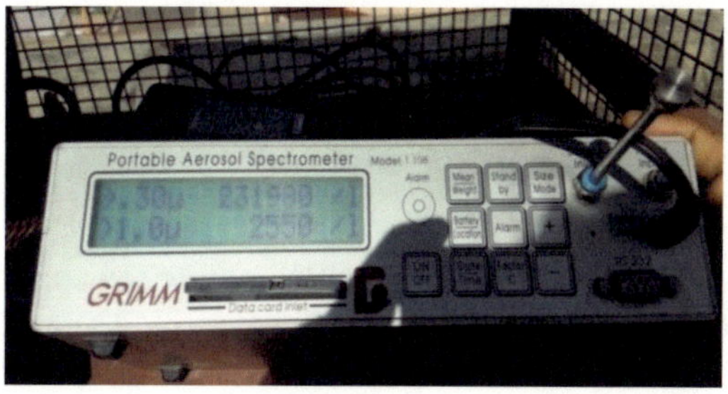

Figure 4. GRIMM aerosol spectrometer (model 1.108).

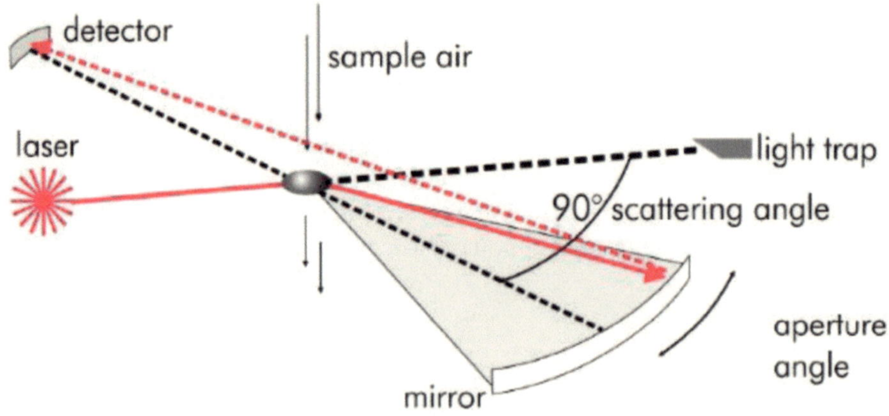

Figure 5. Measuring principle of GRIMM aerosol spectrometer [23].

light pulse of every single particle is being counted and the intensity of its scattering light signal classified to a certain particle size. The measuring principle is schematically shown in **Figure 5** [22].

The spectrometer uses a light-scattering technology for single-particle counts, whereby a semiconductor-laser is used as the light source. The wavelength for laser light is in the infrared ranges, i.e., 780 nm. The aerosol particles can be detected over a very wide size range from 0.3 up to 20 μm. The scattered light is picked up directly as well as via a specific wide-angle optical system with a reflector at a scattering angle of 90°. The signal of the detector is classified into size channels after amplification subject to its intensity. This way, the dependency of the refraction index on the measured signal is low and scattered light intensity permits a precise determination of the particle size. The measured particles are deposited on a removable 47 mm polytetrafluoroethylene (PTFE) filter inside the instrument and are thus available for further gravimetric analysis as well as chemical analysis for detection of metals and nonmetals, etc. **Figure 5** shows the assembly of the laser-measuring chamber. The sample air duct occurs perpendicular to the perspective into the measuring volume [23]. The advantages of ground-based instruments are better accuracy and precise measurements. However, the main disadvantage is that data are sparsely available due to highly expensive instruments. Also, one problem which the scientific community faces is data sharing. The benefits of data sharing are improving our understanding of the results or analysis, the accuracy of the research, can also strength collaborations.

3. Satellite-based remote sensing instruments

The high spatial and temporal resolution of satellite remote sensing data is more valuable in most atmospheric studies. The application of these new technologies to different satellite data have led to the generation of multiple aerosol products, such as spatial distribution, temporal variation, a fraction of fine and coarse modes, vertical distribution, light absorption, and some spectral characteristics. These can be used to infer sources of major aerosol emissions, the transportation of aerosols, interactions between aerosols and energy and water cycles, and the involvement of aerosols with the dynamic system. A series of space missions with instruments capable of making observations of backscattered solar radiation emerging at the top of the atmosphere (TOA) have been launched since the late 1970s.

The current satellites orbiting the Earth are either in geostationary or polar orbits and are described below.

1. Geostationary orbits

   Geostationary satellites are orbiting the Earth in the equatorial plane at a distance of about 36,000 km above the surface. The satellite's rotational velocity is identical to that of the Earth, and this enables the monitoring of dynamic meteorological phenomena such as major dust clouds, volcanic eruptions, smoke plumes or regional pollution events as they evolve throughout the day.

2. Polar orbits

   Polar satellites are orbiting the Earth in a plane that passes through the two poles. The orbit is adjusted to about 100 km above the Earth's surface and this corresponds to a full rotation every 90 minutes. Due to the low orbit, each consecutive rotation samples only a swath of the Earth's surface that is at most 3000 km wide. Each swath is incremented such that the polar orbit is sun-synchronous, i.e., the sensor will always observe the sunny side of the Earth at the same local time of the day. But this limits the polar-orbiting satellite to only one measurement or observation per day over any given location on the Earth's surface. The low orbit allows sensors to have higher spatial resolution and geolocation accuracy. Most quantitative aerosol datasets and climatologies are derived from polar orbiting satellites carrying a variety of sensors. There are two basic observation modes for satellite instruments depending on the observation geometry, i.e., vertical and horizontal observation modes (**Figure 6**). In vertical (or nadir viewing) observation mode, the satellite instrument faces to nadir or near-nadir to detect and measure the radiation coming from the Earth. A number of satellite instruments use the vertical observation concept to provide column integrated products. Observations in a horizontal direction including Limb-viewing and occultation sounding probe the Earth's limb at various depths in the atmosphere. Horizontal observations are characterized by altitude and geo-location of the tangent point. Sun occultation instruments can retrieve aerosol extinction profile from measurements of solar extinction through the atmospheric limb during sunrise and sunset.

Some of the polar orbiting satellites providing long-term and continuous sets of aerosol data are described below:

3.1. The ozone monitoring instrument (OMI)

Ozone monitoring instrument (OMI) onboard the NASA's Aura spacecraft has a high spectral resolution UV and visible (270–500 nm) spectrum. OMI is a nadir-viewing, wide-field-imaging UV and visible spectrometer designed to monitor ozone and other atmospheric species

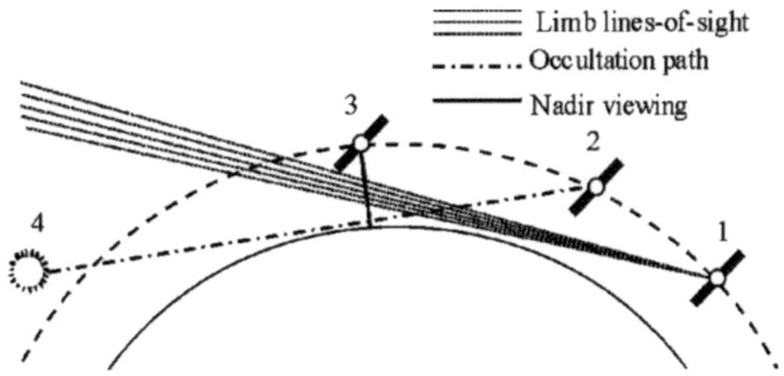

Figure 6. Observation modes of different satellites (1) nadir (vertical), (2) limb, (3) occultation (horizontal), and (4) Sun [24].

including aerosols. It derives its heritage from the descendant TOMS and the European Space Agency (ESA) Global Ozone Monitoring Experiment (GOME) instrument (on the ERS-2 satellite). This instrument has the capability to categorize aerosol types and also measures cloud pressure and coverage. The sensor provides data to derive tropospheric ozone and other important parameters related to climate and ozone chemistry. It uses a push broom technique to measure backscatter in the visible and UV spectrum and has a self-calibration mechanism. The measurements are accurate and precise of total ozone concentration due to its hyperspectral capabilities. OMI instrument has telescope field of view (FOV) 115° wide in across-track dimension. The orbital inclination of Aura Satellite is 98.1 degrees, providing latitudinal coverage from 82° N to 82° S.

3.2. Multi-angle imaging spectroradiometer (MISR)

The multi-angle imaging spectroradiometer is a satellite sensor onboard the NASA's Terra satellite. This scientific instrument was designed to measure the intensities of reflected and absorbed solar radiations by Earth in various directions and spectral bands. It has nine different digital cameras each having four spectral bands (blue, green, red, near infrared), and views the Earth at nine different angles. Out of these 9 cameras, one is towards nadir, other cameras as tilted forward and afterward view angles, at the Earth's surface, of 26.1, 45.6, 60.0, and 70.5°. When it is over the Earth surface each region is imaged by all nine cameras in four spectral bands. It can measure the amount and type of aerosol particles produced due to natural and anthropogenic activities. It can also measure the type and height of clouds. In addition, it also can distinguish different land surface types (snow and ice fields, vegetation, etc.). Owing to nine different angles it covers equator coverage in 9 days and Polar coverage in 2 days.

3.3. Moderate resolution imaging spectroradiometer (MODIS)

This is a key instrument onboard the Terra and Aqua satellite. Terra satellite orbits around the Earth during the morning at 10:30 local time in the north to south direction (descending) while Aqua orbits during the afternoon at 13:30 local time in the south to the north direction (ascending). Both of these instruments capture the same area on Earth. Though, they are approximately 3 hours apart. They observe the earth surface every 1–2 days. MODIS measures

http://dx.doi.org/10.5772/intechopen.80489

various aerosol optical properties, e.g., AOT, types of aerosols, aerosol size-distribution which enable us to understand complex atmospheric processes and varying climate dynamics over land and oceans. Different algorithms are used nowadays for the dark surface as well as bright surface (deep blue). It is providing a long-term and continuous measurement of different aerosol properties from a very long time. This sensor is playing an important role in validating and developing new models to predict climate change. Out of 36 spectral bands, first 19 are used for land, cloud, aerosol (properties and boundaries) as well as for ocean color, atmospheric water vapor, and biogeochemistry, etc., and the remaining are used for surface, cloud, atmospheric (temperature, altitude) as well as cloud water vapor, etc.

Information regarding the general characteristics (Platform, Resolution, Swath Width, Channels, Launch Date, etc.) of OMI, MISR and MODIS are provided in **Table 1**. The list of websites providing design specifications, principles of satellite instruments used for aerosol retrieval can be found elsewhere [25–27]. MODIS and MISR work on Rayleigh scattering principle whereas OMI works on both Rayleigh as well as Mie scattering principles. Both MODIS and OMI have wide spatial coverage (2330 and 2600 km) whereas MISR has narrow coverage (380 km). MODIS and OMI are single view sensors whereas MISR is multi-view sensor (nine cameras). Advantages of all these satellite sensors are wide spatial and temporal coverage. The disadvantage is only one or two retrievals possible in a day.

Sensor	OMI	MISR	MODIS
Platform	NASA EOS-AURA	NASA Terra	NASA Terra/Aqua
Spatial resolution at Nadir (km)	13 km × 24 km at nadir for UV2 (307–383 nm) and visible spectrum (349–504 nm) 13 km × 48 km for UV-1 spectrum (264–311 nm)	0.275	0.25 (bands 1–2) 0.50 (bands 3–7) 1.0 (bands 8–36) Pixel resolution 10 km
Swath width (km)	2600	380	2330
Channel and spectral range(wavelength nm)	270–500 nm with spectral resolution of 0.5 nm	Four wavelengths in each camera Centered at 446, 558, 672, and 867 nm Giving image in red, green, blue and near infrared	620–965 nm and 3.6–14.385 μm 36 spectral bands
Launch	15th July 2004	18th December 1999	18th December 1999/4th May 2002
Equator crossing time and mode	13:42 (ascending mode)	10:30 (descending mode)	10:30 (descending mode)/ 13:30 (ascending mode)
Multiview observations	No	9 (yes)	No
Altitude (km)	705	705	705

Table 1. General characteristics of instruments currently used for aerosol retrievals.

4. Conclusions

In this chapter, we discussed the principles and use of ground and satellite-based remote sensing instruments like CIMEL sunphotometer, MICROTOP, mass spectrometer. The remote sensing observations are commonly used to monitor and study the interaction of aerosols with solar radiation by measuring several aerosol optical properties on temporal and spatial variations on both local and global scales. Satellite based remote sensing owing to wide spatial and temporal coverage enables us to get retrievals or measurements at every spatial point. Sensors like OMI, MODIS, MISR have provided us long-term and continuous measurements of AOD and other selected optical parameters at every location with some bias factor. Although these instruments are sparsely available and also very expensive, these instruments yield reasonably good quality data, along with aerosol spectra, with a valid radiometric calibration. The vertical column aerosol load and aerosol types can be computed on basis of the aerosol spectra. These sensors have advantages and disadvantages though new techniques developed nowadays have enabled us to estimate air quality relatively better using a combination of satellite retrievals along with chemical transport model outputs.

Acknowledgements

Part of the work was performed while Sunita Verma (SV) was working at Birla Institute of Technology (BIT) Mesra, Jaipur Campus, India. SV thanks the funding support from MOES project (MOES/16/18/2017-RDEAS) and BIT administration especially executive director BISR, Prof. P Ghosh for providing time and resources. KR thanks funding support from DST under Inspire Faculty award (IFA-12-EAS-02) and SERB project (ECR-000490). Authors also like to thank AERONET, MODIS mission scientists and authorized personnel associated with the linked projects.

Author details

Sunita Verma[1*], Divya Prakash[2], Manish Soni[3] and Kirpa Ram[1]

*Address all correspondence to: verma.sunita@gmail.com

1 Department of Environment and Sustainable Development, Banaras Hindu University, Varanasi, Uttar Pradesh, India

2 Department of Civil Engineering, Poornima University, Jaipur, Rajasthan, India

3 Department of Physics, Birla Institute of Technology Mesra, Jaipur, Rajasthan, India

References

[1] Charlson RJ, Schwartz SE, Hales JM, Cess RD, Coakley JA Jr, Hansen JE, Hofmann DJ. Climate forcing by anthropogenic aerosols. Science. 1992;**255**:423-430

[2] Hansen J, Sato M, Lacis A, Ruedy R. The missing climate forcing. Philosophical Transactions of the Royal Society B: Biological Sciences. 1997;**352**(1350):231-240

[3] Kaufman YJ, Tanré D, Boucher O. A satellite view of aerosols in the climate system. Nature. 2002;**419**(6903):215

[4] Pandis SN. In: McMurry PH, editor. Atmospheric Aerosol Properties, in Particulate Matter Science for Policy Makers. Cambridge: Cambridge University Press; 2004

[5] https://aeronet.gsfc.nasa.gov/

[6] https://giovanni.gsfc.nasa.gov/giovanni/

[7] https://ladsweb.modaps.eosdis.nasa.gov/search/

[8] Holben BN, Tanre D, Smirnov A, Eck TF, Slutsker I, Abuhassan N, Newcomb WW, Schafer JS, Chatenet B, Lavenu F, Kaufman YJ. An emerging ground-based aerosol climatology: Aerosol optical depth from AERONET. Journal of Geophysical Research: Atmospheres. 2001 Jun 16;**106**(D11):12067-12097

[9] Solar Light Co. User's Guide of MICROTOPS II. Version 5.5. Philadelphia, USA; 2003. pp. 5-15

[10] https://solarlight.com/product/microtops-ii-sunphotometer/

[11] Ganesh KE, Umesh TK, Narasimhamurthy B. Site specific aerosol optical thickness characteristics over Mysore. Aerosol and Air Quality Research. 2008;**8**(3):295-307

[12] http://solarlight.com/wp-content/uploads/2014/05/MTPMAN-RevC.pdf

[13] Young AT. Air mass and refraction. Applied Optics. 1994;**33**(6):1108-1110

[14] Morys M, Mims FM, Hagerup S, Anderson SE, Baker A, Kia J, Walkup T. Design, calibration, and performance of MICROTOPS II handheld ozone monitor and sun photometer. Journal of Geophysical Research: Atmospheres. 2001;**106**(D13):14573-14582

[15] https://www.cimel.fr/?instrument=sun-sky-lunar-multiband-photometer&lang=en

[16] Holben BN, Eck TF, Slutsker I, Tanre D, Buis JP, Setzer A, Vermote E, Reagan JA, Kaufman YJ, Nakajima T, Lavenu F. AERONET—A federated instrument network and data archive for aerosol characterization. Remote Sensing of Environment. 1998;**66**(1):1-6

[17] Shaw GE. Error analysis of multi-wavelength sun photometry. Pure and Applied Geophysics. 1976;**114**(1):1-4

[18] Box MA, Deepak A. Atmospheric scattering corrections to solar radiometry. Applied Optics. 1979;**18**(12):1941-1949

[19] Russell PB, Livingston JM, Dutton EG, Pueschel RF, Reagan JA, Defoor TE, Box MA, Allen D, Pilewskie P, Herman BM, Kinne SA. Pinatubo and pre-Pinatubo optical-depth spectra: Mauna Loa measurements, comparisons, inferred particle size distributions, radiative effects, and relationship to lidar data. Journal of Geophysical Research: Atmospheres. 1993;**98**(D12):22969-22985

[20] http://www.grimm-aerosol.com/products/indoor-air-quality-monitors/index.php

[21] Mohan M, Payra S. Relation between fog formation and accumulation mode aerosols in urban environment of megacity Delhi. Indian Journal of Environmental Protection. 2006;**26**(4):294

[22] http://www.wmo-gaw-wcc-aerosol-physics.org/files/opc-grimm-model--1.108-and-1.109.pdf

[23] http://cires1.colorado.edu/jimenez-group/Manuals/Grimm_OPC_Manual.pdf

[24] Feofilov AG, Kutepov AA. Infrared radiation in the mesosphere and lower thermosphere: Energetic effects and remote sensing. Surveys in geophysics. 2012 Nov 1;**33**(6):1231-1280

[25] OMI: I, https://aura.gsfc.nasa.gov/omi.html; II, https://docserver.gesdisc.eosdis.nasa.gov/public/project/OMI/README.OMI_DUG.pdf; III, https://disc.gsfc.nasa.gov/datasets/OMO3PR_V003/summary; IV, https://www.nasa.gov/mission_pages/aura/spacecraft/omi.html

[26] MISR: I, https://terra.nasa.gov/about/terra-instruments/misr; II, https://www-misr.jpl.nasa.gov/; III, https://en.wikipedia.org/wiki/Multi-angle_imaging_spectroradiometer

[27] MODIS: I, https://modis.gsfc.nasa.gov/about/; II, https://modis.gsfc.nasa.gov/about/specifications.php; III, https://nsidc.org/data/modis/terra_aqua_differences

Section 3

Risk Assessment

Chapter 5

Extreme Value Analysis and Risk Communication for a Changing Climate

Meng Gao

Additional information is available at the end of the chapter

http://dx.doi.org/10.5772/intechopen.79301

Abstract

Due to climate change, the common assumption of stationarity in extreme value analysis of climate extremes has been gradually challenged. The familiar concepts such as a return period and a return level do not apply in a changing climate. To quantify and communicate risk of climate extremes in a changing climate, nonstationarity should be considered carefully. In this chapter, both the concepts and interpretations of return period, return level, failure risk, and reliability under nonstationary condition were interpreted. It was concluded that the two interpretations of the return period became divergent under nonstationary condition, while the two interpretations of failure risk were consistent irrespective of stationarity. Moreover, two examples of risk communication based on generalized extreme value (GEV) distribution for nonstationary climate extremes were presented. In the first example, climate change and its impacts on global air temperature extremes were detected. In the second example, extreme value analysis was firstly applied to precipitation extremes at two weather stations in China. Then, the fitted GEV distribution on historical records was also extrapolated for future risk communication. With these examples, the concepts those were related to risk measure and communication in a changing climate could be easily understood and applied in practice.

Keywords: extreme value theory, nonstationarity, engineering design, return level, failure risk, reliability

1. Introduction

Extreme climate events could, directly or indirectly, impact all sectors of the economy leading to severe losses of life and property [1–3]. Mitigating natural hazards caused by extreme climate events is crucial to the sustainable development of human society and economy [4]. In IPCC's report, an extreme climate event is generally defined as the occurrence of a value of a weather or climate variable above (or below) a threshold value near the upper (or lower) ends of the range of observed values of the variable [1]. The fundamental probability theory of extreme values has been well developed for a long time and already applied in resolving the practical problems in engineering design and risk management [5–7]. For the most part, extreme value theory (EVT) assumes that extreme events are stationary, and these extremes could be successfully characterized by the probability distributions such as the generalized extreme value (GEV) and generalized Pareto (GP) distribution [8, 9]. The occurrences of extreme events are also assumed to be independent or weakly dependent, then, the return levels and return periods could be easily determined [7, 10]. Under stationary condition, there is a simple one-to-one relationship between a return period and a return level, and these two terms can be easily understood [10, 11]. Moreover, risk could be simply communicated using the probability distributions of extremes derived from EVT [11]. In this study, the term "risk" merely refers to the probability of an extreme event with substantial consequence but not the expected loss in general risk analysis.

During the past few decades, there were clear and convincing evidences for global warming and climate change [12], which raised fundamental interdisciplinary issues of risk analysis and communication [13, 14]. As climate changes, weather or climate extremes also change [4] and gradually challenge the stationarity assumption in climate and weather extreme value analysis [6, 11, 15–17]. It has been documented that in some places, climatic and hydrological extremes exhibit some type of nonstationarity in the form of trends, shifts or a combination of them [18–20]. In a nonstationary world, both the severity and frequency of climate and weather extremes will change [2, 3, 10, 16, 17]. Consequently, extreme value analysis of climate and weather extremes has to consider and account for the nonstationarity [15]. Katz et al. [21] presented a nonstationary GEV distribution by introducing time as a covariate. He further showed that both GEV and GPD distributions could be retained under nonstationarity, and maximum likelihood method was also applicable for parameter estimation [6]. Nonstationary extreme values modeling based on GEV and GPD distributions has been realized in R-package *ismev* [22] and *extRemes* [23]. The other R-package GAMLSS (generalized additive model in location, scale, and shape) also allows nonstationary modeling for block maxima, where the parameters are modeled as linear or smooth functions of covariates [24]. Another available R-package for nonstationary extreme value analysis is *GEVcdn*, in which the parameters in GEV distribution are specified as a function of covariate using a conditional density network [25]. Besides these R-packages, nonstationary extreme value analysis could also been implemented using a MATLAB toolbox *NEVA* [26]. Although the nonstationary models in these packages performed better than the stationary equivalents in fitting nonstationary climate extreme, the return period (or return level) and risk for nonstationary conditions were not explicitly presented.

In nonstationary extreme value analysis, the concepts of the return period (or return level) and risk needed to be carefully reformulated and extended, because these familiar concepts, strictly speaking, no longer apply in a nonstationary climate [27]. In stationary cases, there exist two parallel interpretations for the return period: expected waiting time to an extreme event and expected number of extreme events in a given return period [7, 11]. In addition, the return level is the same in each year under stationary conditions. In Wigley [16], the return period was defined as the expected waiting time, and the influence of nonstationarity on the risk of

http://dx.doi.org/10.5772/intechopen.79301

extremes was presented using some simple probability arguments. The concept of the return period was further extended to nonstationary condition in Olsen et al. [10], where the return period was defined as the expected waiting time until an exceedance as the measure of risk. The alternative interpretation of the return period (expected number of extreme events in a given return period) for nonstationary conditions was clearly explained in Parey et al. [28, 29]. Recently, Cooley [11] reviewed these two definitions of the return period suggested by Olsen et al. [10] and Parey et al. [28, 29], and proposed that the return period could be used to communicate risk in nonstationary climate. From the perspective of engineering design, Salas and Obeysekera [7] illustrated the estimation of the return period and examined the failure risk of hydrological structures in nonstationary climate. Rootzén and Katz [30] also concerned the failure risk in the design period and proposed a risk-based engineering design concept, Design Life Level, which served as the basis of risk communication in a nonstationary climate. In the above literatures, the concepts of the return period or return level have been extended and adapted to nonstationary condition; however, the interrelations between return period and risk communication, especially for engineering design purpose, was still ambiguous. The major reason causing such ambiguity is the diversified explanations of one terminology for different purposes. Therefore, a comprehensive interpretation of the return period (or return level) and failure risk (or reliability) under either stationary or nonstationary conditions simultaneously are needed.

The aim of this chapter is to present the extension process of return period, return level, and failure risk from stationary condition to nonstationary condition in a different way so that the commonness and difference could be clearly identified. Consistent with the way how a return period is defined and derived in some previous literatures, extreme value analysis will apply to the time series of annual maxima in this study. Accordingly, GEV distribution is used to illustrate the computation of the return period (return level) and failure risk (reliability) in nonstationary climate.

2. Concepts and interpretations

In some previous literatures, the interpretation of the return period usually began with the simple one-to-one relationship between a return period and a return level under stationary condition [7, 10, 28–30]. Like Cooley [11], we define random variable M_y as the annual maxima of climate or weather events for year y, and $\{M_y\}$ are assumed to be temporally independent. The cumulative probability distribution of M_y is denoted by

$$F_y(x) = P(M_y \leq x) \tag{1}$$

In this study, we try to explain the concepts of the return period and failure risk using the time series of annual maxima and the underlying stochastic process but omitting the assumption of stationarity or nonstationarity. In practice, analyzing the extremes and their probability distributions is usually considered as the basis of frequency analysis, engineering design, and risk assessment.

2.1. Waiting time-based concepts

Given a exceedance level x, let T be the waiting time (from $y = 0$) until an exceedance over this level x occurs [11], then the discrete probability density of random variable T is generally given by [7, 11]:

$$\begin{aligned} P(T=t) &= P\left(M_1 \le x, M_2 \le x, \cdots, M_{(t-1)} \le x, M_t > x\right) \\ &= P(M_1 \le x)P(M_2 \le x)\cdots P\left(M_{(t-1)} \le x\right)P(M_t > x) \\ &= \prod_{y=1}^{t-1} F_y(x)(1 - F_t(x)) \end{aligned} \tag{2}$$

where the second line in Eq. (2) is based on the temporal independence assumption. Then, the expectation of waiting time T is computed as

$$\begin{aligned} E[T] &= \sum_{t=1}^{\infty} tP(T=t) \\ &= \sum_{t=1}^{\infty} t \prod_{y=1}^{t-1} F_y(x)(1 - F_t(x)) \\ &= 1 + \sum_{i=1}^{\infty} \prod_{y=1}^{i} F_y(x) \end{aligned} \tag{3}$$

The details of the derivations of Eq. (3) were shown in the appendix in [11]. The first definition of the return period is based on the expected waiting time. Specifically, a Y-year return period can be interpreted as: the expected time to the next extreme event is Y years [10].

Next, we adopt the commonly used definition of failure risk for an engineering structure, which is interpreted as the probability of the failure or the probability of exceedance over its design level in its design life period. We denote the failure risk by R and the design life period by L (in frequency analysis or engineering design, the denotations L and Y were usually not strictly distinguished). In terms of expected waiting time, the failure risk of a focal structure within its design life period is equivalent to the probability that the expected time of exceedance is less than or equals to the length of the design period, $R = P(T \le L)$. Accordingly, the non-exceedance probability $P(T \le L)$ can also be given by the cumulative probability of the waiting time T [7]:

$$\begin{aligned} R &= P(T \le L) \\ &= \sum_{t=1}^{L} P(T=t) \\ &= \sum_{t=1}^{L} \prod_{y=1}^{t-1} F_y(x)(1 - F_t(x)) \\ &= 1 - \prod_{y=1}^{L} F_y(x) \end{aligned} \tag{4}$$

Consequently, the reliability of the focal structure within its design life period is $R_\ell = 1 - R$.

2.2. Expected number-based concepts

We define random variable N as the number of exceedances over a given exceedance level x occurring in Y years period beginning with the year $y = 1$ and ending with the year $y = Y$ [7, 11]. In each year, we have the following indictor function:

$$I(M_y > x) = \begin{cases} 1, & M_y > x \\ 0, & M_y \le x \end{cases} \tag{5}$$

Then, we get

$$N = \sum_{y=1}^{Y} I(M_y > x) \tag{6}$$

The expectation of N becomes

$$\begin{aligned} E[N] &= \sum_{y=1}^{Y} E[I(M_y > x)] \\ &= \sum_{y=1}^{Y} P(M_y > x) \\ &= \sum_{y=1}^{Y} (1 - F_y(x)) \end{aligned} \tag{7}$$

Now, we say that the Y-year return period can also be interpreted in an alternative way: in Y years the expected number of exceedance events is 1 [28, 29].

Similarly, the reliability of a focal structure in its design life period L can be understood as there are no exceedance events occurring from $y = 1$ to $y = L$. Then, the reliability can be computed as

$$\begin{aligned} R_\ell &= P(M_1 \le x, M_2 \le x, \cdots, M_{(L-1)} \le x, M_L \le x) \\ &= \prod_{y=1}^{L} P(M_y \le x) = \prod_{y=1}^{L} F_y(x) \end{aligned} \tag{8}$$

From Eqs. (4) and (8), we find that the two parallel interpretations of failure risk (or reliability) of a focal engineering structure are equivalent irrespective of $\{M_y\}$ is stationary or nonstationary.

3. Risk communication

3.1. Risk measure under stationarity

Under a stationary assumption, $\{M_y\}$ is identically distributed with a distribution function $F(x)$, where the year index y is discarded for notational simplicity. Now, the relationship between a return period (Y) and the associated return level (x_Y, a special exceedance level) can be revealed by the following equation [11, 30]:

$$F(x_Y) = P(M \le x_Y) = 1 - 1/Y \tag{9}$$

The Y-year return level of annual extreme M is defined to be the $(1 - 1/T)$-th quantile of the distribution of climate extreme in any year. In addition, we have $P(M > x_Y) = 1/Y$. That means that the exceedance probability over the return level x_Y is $1/Y$ for each year.

It has been proved that the two interpretations of return period in the stationary case are both correct with this identical exceedance probability under stationarity assumption [11]. Substituting Eq. (9) into Eq. (3), we get the interpretation of the return period based on waiting time of exceedance:

$$E[T] = 1 + \sum_{i=1}^{Y} \prod_{y=1}^{i} (1 - 1/Y) = Y \tag{10}$$

Similarly, substituting Eq. (9) into Eq. (7), we get the alternative interpretation of return period based on expected number of exceedance events:

$$E[N] = \sum_{i=1}^{Y} 1/Y = 1 \tag{11}$$

The simple one-to-one relationship between a return period and a return level in the stationary case has been commonly utilized in frequency analysis and engineering design practice [8, 21]. For example, the frequency or expected waiting time of extreme events exceeding a given exceedance level can be easily determined using Eq. (9) in frequency analysis of climate extremes. In practice, a very important concept for an engineering structure is the design life period. Reversely, given a design life period or exceedance probability, return levels could also be determined easily. Moreover, the failure risk or reliability of a focal structure in its design life period L could also be evaluated using a simpler formulation

$$R = 1 - (F(x_D))^L \tag{12}$$

where x_D is the design level in engineering design.

3.2. Risk measure under nonstationarity

Under nonstationary condition, $\{M_y\}$ is no more identically distributed. In frequency analysis, engineering design, and risk assessment, the dependence of probability distributions $F_y(x)$ on the year index y should be considered. It is more valuable to do extreme value analysis within the design life period. We have shown the two different interpretations of a return period in Section 2. Under nonstationary condition, the relationship between the return period and the associated return level could be expressed independently using Eqs. (3) and (7). Given a return period or design life period Y (Y and L are substitutable here), the Y-year return level could be estimated by setting $E[T] = Y$ and $E[N] = 1$, respectively [10, 28, 29]. Theoretically speaking, the

Y-year return level in the nonstationary case could be estimated by solving the following two equations numerically

$$Y = 1 + \sum_{i=1}^{\infty} \prod_{y=1}^{i} F_y(x_Y) \tag{13}$$

$$1 = \sum_{i=1}^{Y} \left(1 - F_y(x_Y)\right) \tag{14}$$

To determine $F_y(x)$, fitting the historical records of annual maxima to nonstationary extreme value distribution is the first step. Moreover, to estimate the return level of extremes or assessing the failure risk of a focal structure in its rest life span, it is necessary to extrapolate the trend or shift in climate extremes. Cooley [11] showed that it was unnecessary to extrapolate $F_y(x)$ indefinitely and an accurate estimation of the return level could be obtained, when $F_y(x)$ was monotonically increasing. For computational simplicity, the definition of the return period based on the expected number of events has more advantage since the maximum extrapolation length is Y years but not indefinitely to $+\infty$.

The return level in Eqs. (13) and (14) are the two extensions of the return period in the stationary case; however, these two extensions are not applicable in practical engineering design [7, 30]. For engineering design purpose, Rootzén and Katz [30] presented a new concept, Design Life Level, by keeping the failure risk at a low constant level during the design life period. The relationship between Design Life Level and design life period was expressed by the following equation [30]:

$$\begin{aligned} F_{1\sim Y}(x) &= P(M_{1\sim Y} \leq x) \\ &= P(M_1 \leq x)P(M_2 \leq x)\cdots P\left(M_{(t-1)} \leq x\right)P(M_t \leq x) \\ &= F_1(x) * F_2(x) * \cdots * F_Y(x) \end{aligned} \tag{15}$$

where $M_{1\sim Y} = max\{M_1, M_2, \cdots, M_Y\}$ denoted the largest annual maxima during the design life period $1 \sim Y$. Usually, the mathematical expression of $F_{1\sim Y}(x)$ is analytically intractable, while its numerical approximation $\widehat{F}_{1\sim Y}(x)$ is frequently used in practice. Given a failure risk, $\widehat{r}$, of a focal engineering structure during its design period $1 \sim Y$, the associated design life level could be computed by

$$DLL = \widehat{F}_{1\sim Y}^{-1}(1 - \widehat{r}) \tag{16}$$

A variant of Design Life Level is Minimax Design Life Level [30]. The computation of Minimax Design Life Level is even simpler. During the whole design life period $1 \sim Y$, we can obtain a series of return levels: $\left\{\widehat{F}_y^{-1}(1 - \widehat{r})\right\}, y = 1, 2, \cdots, Y$, and the Minimax Design Life Level is

$$minmaxDLL = max\left\{\widehat{F}_y^{-1}(1 - \widehat{r})\right\}, y = 1, 2, \cdots, Y \tag{17}$$

Similarly, the first step to compute the Design Life Level or Minimax Design Life Level is nonstationary modeling of historical climate extremes. Then, the trends of extremes would be extrapolated over the design life period. Moreover, the statistical uncertainty in the return period and Design Life Level can be described by computing the standard errors using the delta method [11, 30].

4. Applications

In this section, we present two examples of extreme value analysis and risk communication in a changing climate. The cumulative distribution function of the GEV is expressed as [8]:

$$F(x) = \exp\left\{-\left[1+\varepsilon\left(-\frac{x-\mu}{\sigma}\right)\right]^{-1/\varepsilon}\right\}, 1+\frac{\varepsilon(x-\mu)}{\sigma} > 0 \tag{18}$$

where μ, $\sigma > 0$, and ε are the location, scale, and shape parameters, respectively. Constant parameters correspond to stationary GEV distribution, while time-varying parameters correspond to nonstationary GEV distribution. The time-varying parameters in nonstationary GEV distribution could be modeled as the function of time or other climate indictors [6]:

$$F_y(x) = \exp\left\{-\left[1+\varepsilon_y\left(-\frac{x-\mu_y}{\sigma_y}\right)\right]^{-1/\varepsilon_y}\right\} \tag{19}$$

where y is the year index. Commonly, the location parameter μ_y and/or the scale parameter σ_y are assumed to be time varying, while the shape parameter is assumed to be constant [6–8, 26]. In particular, the extrapolation of $F_y(x)$ into the future design life period is reasonable, only if the location and/or the scale parameters have linear or log-linear trends [7, 26]. Before extrapolation, it is needed to select a best fitting GEV distribution model, and the model selection is usually based on AIC or BIC [6].

The first example of risk communication was for global annual maximum near surface air temperature (1948–2015). The global gridded data were extracted from the reanalysis products with a spatial resolution of 2.5 * 2.5 provided by Earth System Research Laboratory, NOAA (http://www.esrl.noaa.gov). For each grid, the time series of annual maximum near surface air temperature from 1948 to 2015 was firstly constructed and the trend was detected using the Mann-Kendall (M-K) test method [32, 33]. The test result was showed in **Figure 1(a)**. Both positive and negative trends at the 5% significance level were detected during the past 68 years (1948–2015) for most part of the earth. The time series with significant trends will be fitted using nonstationary GEV distribution with time-varying parameters. Otherwise, a stationary GEV distribution with constant parameters will be applied. Like Cheng et al. [26], only the location parameter was assumed to be linearly varying with time. Nonstationary modeling was performed with the R-package *extRemes* [23]. The aim of this example was to show the changes in climate extremes caused by climate change and how this change impacted risk communication; therefore, we did not extrapolate the trends of temperature extremes but only computed the 20-year return level (the expected number-based return level during 1996–2010). Solving Eq. (14)

http://dx.doi.org/10.5772/intechopen.79301

relied on numerical optimization techniques, and in this study, the particle swarm optimization method was applied. The result of the global 20-year return level of annual maximum near surface air temperature in 1996–2015 was shown in **Figure 1(b)**.

In the second example, we used two time series of annual maximum precipitation (AMP) to illustrate the risk measure and communication under both stationary and nonstationary conditions. The two AMP time series were extracted from observation dataset of daily precipitation,

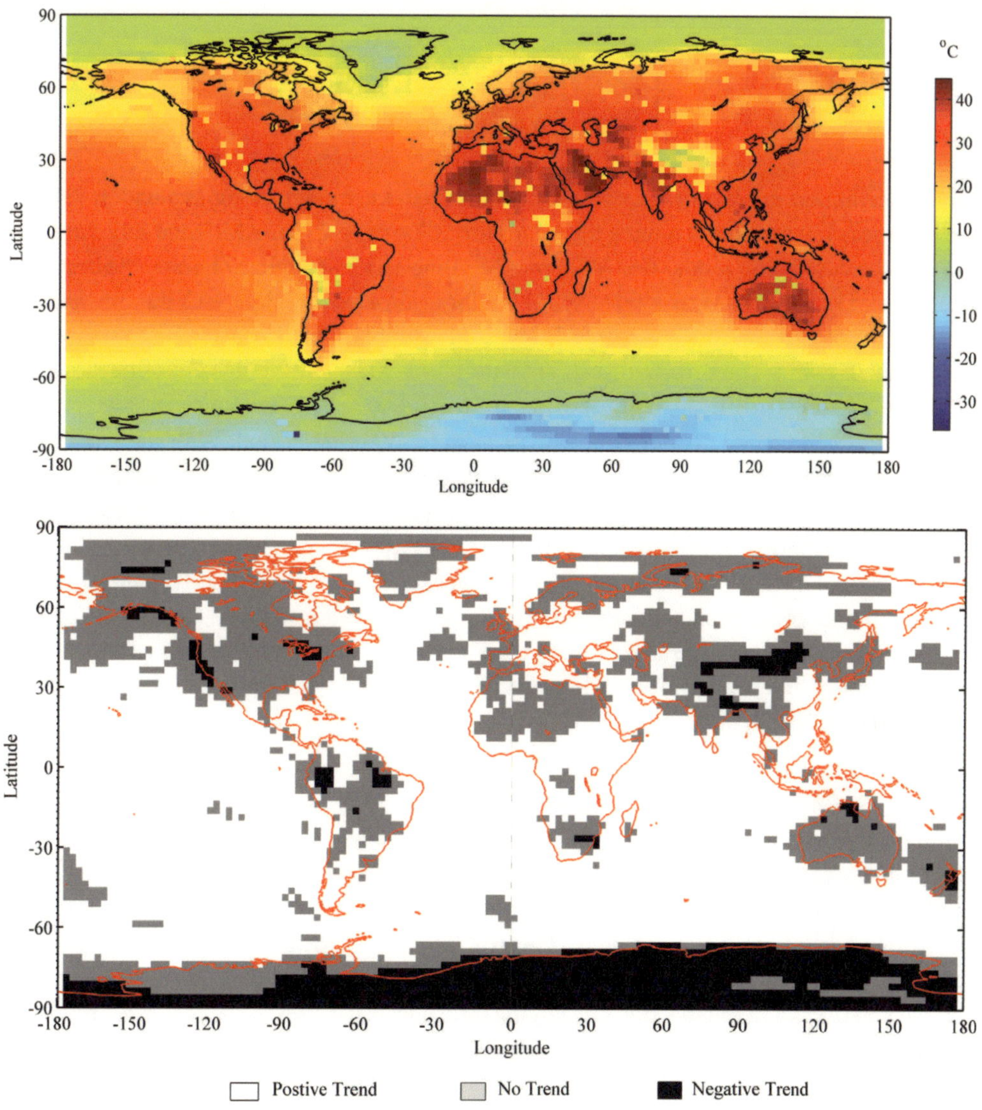

Figure 1. (a) M-K test for global annual maximum near surface air temperature (1948–2015) (positive trend in white: no significant trend in gray; negative trend in black). (b) Nonstationary 20-year return level of global annual maximum near surface air temperature based on the expected number of events during 1996–2015.

which was provided by the National Meteorological Information Center (NMIC) of the China Meteorological Administration (CMA). The two AMP time series were selected, because either positive or negative trends were detected. The corresponding weather stations are Qionghai (Station ID: 59855) and Zunhua (Station ID: 54429), which are located at N19°14′E110°28′ and N40°12′E117°57′, respectively. The valid observation periods were 1953–2013 (Qionghai) and 1956–2013 (Zunhua). Both stationary and nonstationary GEV distributions were used to fit the AMP time series denoted by the following four candidate models:

$$\begin{cases} M0: & \{\mu, \sigma, \varepsilon\} \\ M1: & \{\mu_0 + \mu_1 y, \sigma, \varepsilon\} \\ M2: & \{\mu, \sigma_0 + \sigma_1 y, \varepsilon\} \\ M3: & \{\mu_0 + \mu_1 y, \sigma_0 + \sigma_1 y, \varepsilon\} \end{cases} \tag{20}$$

Akaike information criterion (AIC) was also computed for model fitting evaluation [31]. The model that was preferred was having the minimum value of AIC. For AMP time series at station Qionghai, the best fitting model was $M1$. For AMP time series at station Zunhua, the best fitting model was $M3$. The observed values of AMP, the estimated median, and the 5th and 95th percentiles were shown in **Figure 2(a)**. With the best fitting models, trends in precipitation extremes were extrapolated to the next 50 years (2014–2063). In other words, the design life period was assumed to be 50 years starting from 2014 to 2063. The scale parameter σ_y was constrained to be positive by $max\{0, \sigma_0 + \sigma_1 y\}$ in the design life period. With the extrapolated $F_y(x)$ in 2014–2063, we computed the return levels (or Design Life Level and Minimax Design Life Level) along with their standard errors, expected waiting time (or the return period that has been given in advance), failure risk, and reliability in the design life period. The return levels for nonstationary conditions presented in [10, 28, 29] were computed using Eqs. (13) and (14), while Design Life Level and Minimax Design Life Level are computed using Eqs. (16) and (17), respectively. The corresponding standard errors were computed using the delta method [11, 30]. Expected waiting time was computed based on Eq. (3), and the failure risk and reliability were computed using Eqs. (4) and (8).

The results were shown in **Tables 1** and **2**, respectively. The first three rows in **Tables 1** and **2** mainly illustrated the relationship between the return period and the return level under stationary and nonstationary conditions. The last two rows in **Tables 1** and **2** showed the two concepts, Design Life Level and Minimax Design Life Level, for the purpose of engineering design under nonstationary conditions. For AMP time series at Qionghai station, there was a significant positive trend in precipitation extremes. Given a 50-year return period, the associated return level under stationary assumption was much lower than those under nonstationary assumption. From the perspective of engineering design, the return levels shown in the first three rows were unacceptable, because the failure risks in the following 50 years (design life period) were all larger than 0.55. To ensure a low failure risk, a higher design level is needed. For AMP time series at Zunhua station, there was a significant negative trend in precipitation extremes. When nonstationarity was considered, the return level became lower due to the decreasing trend in precipitation extremes. Furthermore, due to the same reason, the Design Life Level and Minimax Design Life Level in 50-year design life period (2014–2063) were lower than the 50-year return level under stationary assumption.

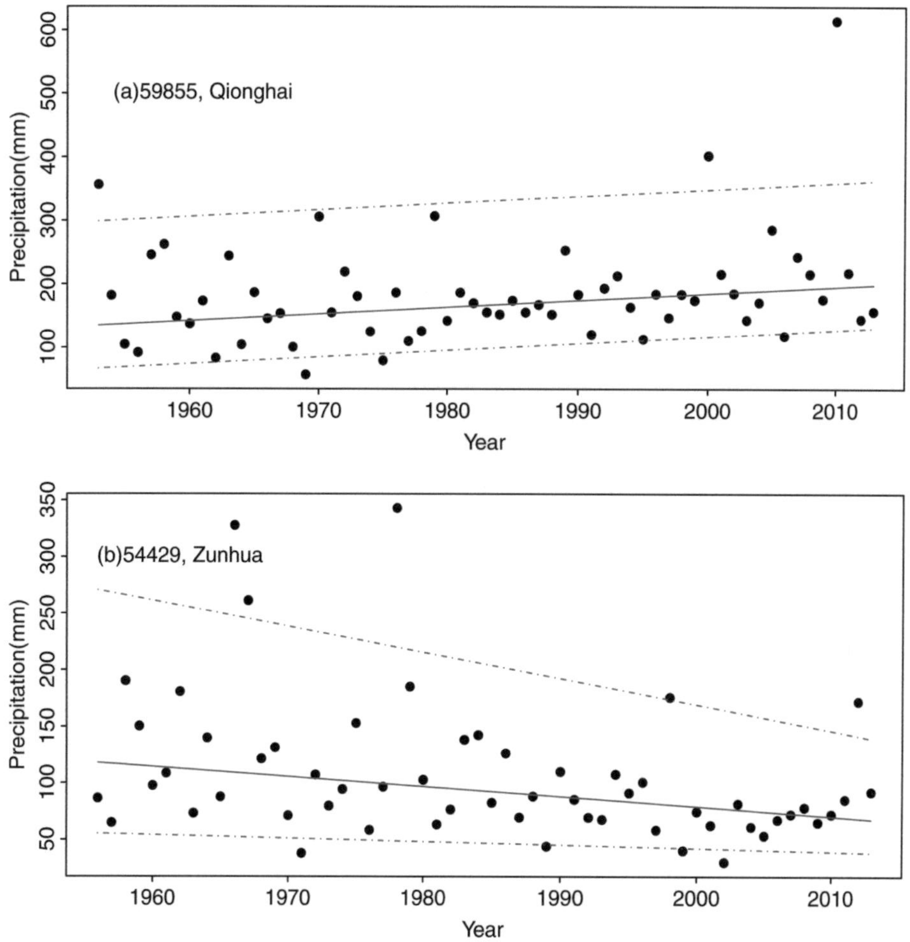

Figure 2. Summary of the nonstationary modeling of annual time series of precipitation extremes using GEV distribution models. (a) Station Qionghai (ID:59855); observation period: 1953–2013; the best fitting nonstationary GEV distribution: *M*1. (b) Station Zunhua (ID:54429); observation period: 1956–2013; the best fitting nonstationary GEV distribution: *M*3. Symbols: observed values (dots), the estimates of the median (solid lines), and the 5th and 95th percentiles (dashed lines).

		Risk communication				
Model	**Equation**	**Return level**	**Standard error**	**Return period (or EWT)[a]**	**Risk**	**Reliability**
*M*0	Eq. (9)	398.31	46.15	50	0.6358	0.3642
	Eq. (13)	485.00	65.67	50	0.5594	0.4406
*M*1	Eq. (14)	466.61	61.77	50	0.6359	0.3641
	Eq. (16)	818.87	13.4	177.03	0.0407	0.9593
	Eq. (17)	790.58	15.0	171.96	0.05	0.95

[a]EWT stands for expected waiting time.

Table 1. Results of risk communication for precipitation extremes with positive trend at station Qionghai (ID: 59855).

		Risk communication				
Model	**Equation**	**Return level**	**Standard error**	**Return period (or EWT)[a]**	**Risk**	**Reliability**
*M*0	Eq. (9)	282.3	50.59	50	0.6358	0.3642
	Eq. (13)	81.44	20.07	50	0.7647	02353
*M*3	Eq. (14)	83.72	18.75	50	0.6548	0.3452
	Eq. (16)	105.62	11	199.78	0.001	0.9989
	Eq. (17)	98.41	19.59	189.98	0.05	0.95

[a]EWT stands for expected waiting time.

Table 2. Results of risk communication for precipitation extremes with negative trend at station Zunhua (ID: 54429).

5. Discussion and conclusions

Due to the climate change, the stationary assumption that was commonly used in statistical analysis of climate extremes gradually became unacceptable [6, 15, 17]. How to quantify and communicate risk of climate extremes in nonstationary climate is essential for engineering design and risk assessment [7, 11, 30]. There were many attempts to quantify and communicate risk in a changing climate such as extending the concepts of the return period from stationary condition to nonstationary condition [10, 28, 29] or developing a new concept of the return level [30]. In stationary climate, frequency analysis, engineering design, and risk assessment were all based on the stationary extreme value distribution model [8]. It was assumed that the fitted extreme value distribution model on historical records also applied for future observations. Also due to stationarity, the concepts of risk measure for different purposes had not been strictly distinguished. Unlike the simple one-to-one relationship between a return level and a return period under stationary condition, risk measure and communication were more complicated under nonstationary condition, especially due to the time-varying essence of climate extremes. Therefore, a clear interpretation and illustration of the methods for risk measure and communication in a changing climate are of great importance.

In this study, climate extremes were presented in the form of annual maxima of extreme climate events. This chapter began with the two parallel interpretations of the return period, in which, the implicit relationship between a return level and a return period was included, but the stationary or nonstationary assumptions were omitted. This implicit relationship was also considered as the basis for frequency analysis and engineering design. In the stationary case, the two interpretations of the return period were equivalent. Although they were no more equivalent in the nonstationary case, they both provided independent methods for determining the associated return level for a given return period. Risk assessment usually aims to a focal engineering structure with a given design level. We showed that the concept of failure risk (or reliability) also had two parallel interpretations, and these two interpretations were consistent irrespective of stationary or nonstationary assumptions. In order to illustrate how risk was quantified and communicated in a changing climate, two examples of nonstationary climate extremes were used. Totally, we have reviewed two methods for estimating the return level for a given return period under nonstationary condition [10, 28, 29] and two newly refined concepts of the return level in a given design life period for engineering design purpose [30]. In the first example, we detected the trend of annual maximum of global near surface air temperature during 1948–2015. Nonstationary GEV distribution with a time-varying location parameter was used to fit near surface air temperature extremes with significant trends, and the expected number-based return levels in 1996–2015 were computed. In the second example, time series of observed annual precipitation extremes at two weather stations in China with significant trends was analyzed. Both stationary and nonstationary GEV distribution models were used to fit the precipitation extremes. For each station, one best fitting GEV distribution was identified. Then, the linear trends in the parameters of GEV distribution were extrapolated into the following 50 years (also considered as the design life period). Return level, Design Live Level, and Minimax Design Live Level were all computed for the design life period (2014–2063), respectively. It was concluded that the communication of risk in a changing climate was obviously different from that in a stationary climate. For frequency analysis purpose, general return level/return period might be quite capable of communicating risk of climate extremes. While for the engineering design purpose, Design Live Level or Minimax Design Live Level were recommended, because the failure risk of a focal structure would be very low if Design Live Level or Minimax Design Live Level were chosen as the design level.

A reliable statistical modeling on long-term data was the basis of risk communication in a changing climate [26]. In nonstationary extreme value modeling, there are usually many candidate models. The choice of extreme value distribution models might influence the risk measure substantially in nonstationary and changing climate, because the trend captured by the extreme value distribution should be extrapolated into the future design life period. In this chapter, we only chose time as the covariate in the nonstationary extreme value analysis, and the parameters in GEV distribution model were expressed as the linear function of time. Perhaps, there might be more suitable trends such as quadratic or exponential trends leading to more candidate models. Evaluating all these candidate models was not an easy task. In practice, only a few of commonly used models was evaluated and compared. Moreover, the model selection process is not simply by using tools such as AIC or BIC [11]. Sometimes, additional expert knowledge is needed. In nonstationary extreme value modeling, besides time, some other climate indictors representing the variability of the climate system were also chosen as covariates. Although the historical data could be successfully fitted with these additional climate indictors, it was difficult to extrapolate the historical climate variability. That is because the climate variability itself is difficult to predict due to the complicity of the climate system. To reduce the uncertainty in statistical extrapolation, the output of numerical climate models was also used. However, the reasonability of simulation results was constrained by the parameters setting and initial values [30]. Additionally, the standard error of return levels and Design Live Level was all estimated using the delta method. Although standard error is a simple measure to quantify the uncertainty of nonstationary extreme value modeling, the uncertainty cannot be properly reflected using the symmetric confidence interval. Lastly, as pointed in [30], neither Design Life Level nor Minimax Design Life Level could be used as the criteria for realistic engineering design, because more economic and political factors should be considered besides the failure risk and reliability. All the above mentioned things are outside the scope of this chapter; here, our primary objective was to discriminate the concepts and their interpretations of the return period/return level, failure risk/

reliability under stationary and nonstationary conditions, and to illustrate the computations using realistic climate extremes. With these examples, we believed that the concepts those were related to risk measure and communication in a changing climate could be easily understood and applied.

Acknowledgements

This work was partly supported by the Youth Innovation Promotion Association of CAS (2016195), CAS Knowledge Innovation Project (KZCX2-EW-QN209), and National Natural Science Foundation of China (31570423).

Author details

Meng Gao

Address all correspondence to: mgao@yic.ac.cn

Yantai Institute of Coastal Zone Research, Chinese Academy of Sciences, Yantai, China

References

[1] IPCC. Special Report on Managing the Risks of Extreme Events and Disasters to Advance Climate Change Adaptation. A Report of Working Groups I and II of the Intergovernmental Panel on Climate Change. UK: Cambridge University Press; 2012

[2] Mika J. Changes in weather and climate extremes: Phenomenology and empirical approaches. Climatic Change. 2013;**121**:15-26

[3] Monier E, Gao X. Climate change impacts on extreme events in the United States: An uncertainty analysis. Climatic Change. 2015;**131**(1):67-81

[4] Zhang X, Zwiers FW. Statistical indices for the diagnosing and detecting changes in extremes. In: AghaKouchak A, Easterling D, Hsu K, Schubert S, Sorooshian S, editors. Extremes in a Changing Climate. Springer; 2013. pp. 1-14

[5] Leadbetter MR. Extremes and local dependence in stationary sequences. Probability Theory and Related Fields. 1983;**65**(2):291-306

[6] Katz RW. Statistical methods for nonstationary extremes in extremes in a changing climate, 15–38 Easterling D, Hsu K, Schubert S, Sorooshian S, Extremes in a Changing Climate. Berlin: Springer; 2013

[7] Salas JD, Obeysekera J. Revisiting the concepts of return period and risk for nonstationary hydrologic extreme events. Journal of Hydrologic Engineering. 2014;**19**:554-568

[8] Coles S. An Introduction to Statistical Modeling of Extreme Values. Berlin: Springer; 2001

[9] Li Z, Brissette F, Chen J. Finding the most appropriate precipitation probability distribution for stochastic weather generation and hydrological modelling in Nordic watersheds. Hydrological Processes. 2013;**27**(25):3718-3729

[10] Olsen JR, Lambert JH, Haimes YY. Risk of extreme events under nonstationary conditions. Risk Analysis. 1998;**18**(4):497-510

[11] Cooley D. Return periods and return levels under climate change in extremes in a changing climate. In: AghaKouchak A, Easterling D, Hsu K, Schubert S, Sorooshian S, editors. Extremes in a Changing Climate. Berlin: Springer; 2013. pp. 97-114

[12] IPCC. Climate change 2007: The physical science basis. In: Contribution of Working Group I to the Fourth Assessment Report of the Intergovernmental Panel on Climate Change. Cambridge: Cambridge University Press; 2007

[13] Lorenzoni I, Pidgeon NF, O'Connor R. Dangerous climate change: The role for risk research. Risk Analysis. 2005;**25**:1387-1398

[14] Pidgeon N. Climate change risk perception and communication: Addressing a critical moment? Risk Analysis. 2012;**32**:951-956

[15] Milly PCD, Betancount J, Falkenmark M, Hirsch RM, Zbigniew W, Lettermaier DP, Stouffer RJ. Stationarity is dead: Whither water management? Science. 2008;**319**(5863): 573-574

[16] Wigley TML. The effect of climate change on the frequency of absolute extreme events. Climatic Change. 1988;**17**:44-55

[17] Wigley TML. The effect of changing climate on the frequency of absolute extreme events. Climatic Change. 2009;**97**:67-76

[18] Olsen JR, Stedinger JR, Matalas NC, Stakhiv EZ. Climate variability and flood frequency estimation for the Upper Mississippi and Lower Missouri rivers. Journal of the American Water Resources Association. 1999;**35**(6):1509-1523

[19] Kiem AS, Franks SW, Kuczera G. Multi-decadal variability of flood risk. Geophysical Research Letters. 2003;**30**(2):GL015992

[20] Villarini G, Serinaldi F, Smith JA, Krajewski WF. On the stationarity of annual flood peaks in the continental United States during the 20th century. Water Resources Research. 2009; **45**(8):W08417

[21] Katz RW, Parlange MB, Naveau P. Statistics of extremes in hydrology. Advances in Water Resources. 2002;**25**(8):1287-1304

[22] Stephenson AG. ismev: An Introduction to Statistical Modeling of Extreme Values, R package version 1.35 ed. 2010

[23] Gilleland E, Katz RW. New software to analyze how extremes change over time. Eos. 2011; **92**(2):13-14

[24] Stasinopoulos DM, Rigby RA. Generalized additive models for location scale and shape (GAMLSS) in R. Journal of Statistical Software. 2007;**23**(7):1-46

[25] Cannon AJ. EVcdn: An R package for nonstationary extreme value analysis by generalized extreme value conditional density estimation network. Computational Geosciences. 2011;**37**:1532-1533

[26] Cheng L, AghaKouchak A, Gilleland E, Katz RW. Non-stationary extreme value analysis in a changing climate. Climatic Change. 2014;**127**:353-369

[27] Katz RW. Statistics of extremes in climate change. Climatic Change. 2010;**100**:71-76

[28] Parey S, Malek F, Laurent C, Dacunha-Castelle D. Trends and climate evolutions: Statistical approach for very high temperatures in France. Climatic Change. 2007;**81**:331-352

[29] Parey S, Hoang TTH, Dacunha-Castelle D. Different ways to compute temperature return levels in the climate change context. Environmetrics. 2010;**21**:698-718

[30] Rootzen H, Katz RW. Design Life Level: Quantifying risk in a changing climate. Water Resources Research. 2013;**49**:1-13

[31] Akaike H. A new look at the statistical model identification. IEEE Transactions on Automatic Control B. 1974;**19**:716-723

[32] Mann HB. Nonparametric test against trend. Econometrica. 1945;**13**:245-259

[33] Kendall MG. A new measure of rank correlation. Biometrika. 1938;**30**:81-93